Green Energy and Technology

Climate change, environmental impact and the limited natural resources urge scientific research and novel technical solutions. The monograph series Green Energy and Technology serves as a publishing platform for scientific and technological approaches to "green"—i.e. environmentally friendly and sustainable—technologies. While a focus lies on energy and power supply, it also covers "green" solutions in industrial engineering and engineering design. Green Energy and Technology addresses researchers, advanced students, technical consultants as well as decision makers in industries and politics. Hence, the level of presentation spans from instructional to highly technical.

Indexed in Scopus.

More information about this series at http://www.springer.com/series/8059

Tina Kegl · Anita Kovač Kralj ·
Marko Kegl · Breda Kegl

Nanomaterials for Environmental Application

Fuel Additives for Diesel Engines

Springer

Tina Kegl
Faculty of Chemistry
and Chemical Engineering
University of Maribor
Maribor, Slovenia

Anita Kovač Kralj
Faculty of Chemistry
and Chemical Engineering
University of Maribor
Maribor, Slovenia

Marko Kegl
Faculty of Mechanical Engineering
University of Maribor
Maribor, Slovenia

Breda Kegl
Faculty of Mechanical Engineering
University of Maribor
Maribor, Slovenia

ISSN 1865-3529 ISSN 1865-3537 (electronic)
Green Energy and Technology
ISBN 978-3-030-54707-3 ISBN 978-3-030-54708-0 (eBook)
https://doi.org/10.1007/978-3-030-54708-0

This Springer imprint is published by the registered company Springer Nature Switzerland AG
The registered company address is: Gewerbestrasse 11, 6330 Cham, Switzerland

Preface

Nanotechnologies have been steadily delivering new nanomaterials with exciting properties which may significantly differ from those of corresponding bulk materials. These special properties make nanomaterials potentially very attractive and important constituents in many modern technologies and processes. In this context, diesel engines are no exception and, as it looks, nanomaterials might help in making the diesel engine suited better for the ever-stringent environmental requirements.

Diesel engines are excellent machines which nowadays power immense quantities of vehicles, trains, ships, and various machineries. Unfortunately, they come with some problems, mainly related to engine harmful emissions. This is one of the reasons for intensive research related to development of alternatives which would power our transport in the future. As it stands, however, it does not seem likely that an economically and technically viable replacement for a diesel engine will be available in the very near future. Effectively, this means that diesel engines probably still have a long way to go and one should waste no time in trying to make them better and cleaner. In this context, nanomaterials came into focus as potentially highly valuable fuel additives, offering opportunities in improving diesel engine performance and reducing the engine harmful emissions.

This book is a comprehensive review of opportunities, potentials, and associated challenges related to engagement of nanomaterials as fuel additives in diesel engines. The goal of this book is to provide a one-point entry for quick initial orientation or road map drafting for managers, engineers, and researchers either already working on this field or being about to enter it. This is achieved by providing a summarized overview of the most important nanomaterial-related data published recently in connection with:

- diesel engine fuel injection, spray, and combustion,
- diesel engine characteristics,
- synthesis and characterization of various nanomaterials,
- overview of diesel engine base fuels and preparation of nanofuels,
- nanofuel properties and their effects on diesel engine emissions and other engine characteristics, and

- practical viability of nanofuel usage, ranging from technical issues to environmental and health concerns.

Most of the literature covered and referenced was published within the last few years, especially in the period of 2017–2019. The data published in this literature is elaborated into a summarized form by providing useful tables and charts. So, this book provides an excellent starting point and a source of fresh ideas for anyone being at least partially involved into the development of either cleaner diesel engines or nanomaterials for environmental application.

Maribor, Slovenia

Tina Kegl
Anita Kovač Kralj
Marko Kegl
Breda Kegl

Acknowledgements The authors are grateful for the financial support from the Slovenia Research Agency (Ph.D. research fellowship contract No. 1000-18-0552 and core research funding No. P2 0032, P2 0137, and P2 0196).

Contents

Chapter 1
Introduction

Despite substantial efforts on various alternatives, the world's transport is still almost completely powered by internal combustion engines [12]. Among these, diesel engines play a decisive role since they are used to power vehicles, trains, ships, and, beyond this, also a wide range of various machineries [18]. A diesel engine is an excellent machine with extraordinary properties. However, it also comes with various drawbacks, and perhaps the main problems are its harmful emissions, especially nitrogen oxides (NO_x), smoke, particulate matter (PM), carbon monoxide (CO), and unburned hydrocarbons (HC).

To address the harmful emissions' problem while keeping engine power and efficiency within acceptable limits, the following fields are being investigated intensively [14]:

- engine management,
- after-treatment technologies, and
- usage of various alternative fuels.

Engine management is mostly related to the control of the fuel injection process. Modern electronically controlled fuel injection systems (EC-FIS) must enable high injection pressure capability and injection pressure control, flexible injection timing control, and injection rate control according to the current engine operating regime [14]. In this way, it is possible to mitigate the formation of harmful emissions to a great extent. After-treatment technologies are used to reduce the quantity of pollutants that enter the environment. After-treatment technologies for diesel engines comprise diesel oxidation catalyst (DOC), lean-NO_x trap (LNT), selective catalytic reduction (SCR), diesel particulate filters (DPF), and exhaust gas recirculation (EGR) [5, 9]. DOC is usually used to reduce CO and unburned HC emissions, while LNT and SCR technologies are used mainly to reduce NO_x emission. A DPF can be used to remove PM from diesel engine exhaust gases, while EGR can be engaged to reduce NO_x emissions and improve brake thermal efficiency [5]. Finally, engagement of alternative fuels also offers an opportunity to achieve a reduction of harmful emissions. A wide range of various alternative fuels, either pure or blends from

T. Kegl et al., *Nanomaterials for Environmental Application*, Green Energy and Technology, https://doi.org/10.1007/978-3-030-54708-0_1

various sources, come with rather diverse fuel properties. Therefore, alternative fuel usage typically has to be accompanied by adequately modified engine management and properly adapted after-treatment systems [11].

Nowadays, the most frequently investigated alternative fuel candidates for diesel engines are biodiesels from various source materials, vegetable oils, water-in-diesel emulsion, natural gas and liquefied petroleum gas, methane and propane, dimethyl ether (DME) and dimethyl carbonate (DMC), Fischer–Tropsch diesel, hydrogen, methanol, bioethanol, propanol, and butanol [2, 4, 7, 13, 14, 16, 19]. Some of those alternative fuels can be used pure, while others have to be mixed with mineral diesel or with other alternative fuels.

Among all alternative fuels, biodiesels, alcohols, natural gas, and DME are considered to be four of the most promising and attractive alternatives, because they may easily be acquired, handled, and stored [8]. Among these, biodiesels look to be especially attractive since they can be easily extracted from biomass while being biodegradable and environmentally friendly. Furthermore, they can decrease net CO_2 emissions, air pollution, and reduce acid rain. Additional benefits include sustainability, fuel security, regional development, and a decrease in rural poverty [1].

Unfortunately, the usage of alternative fuels in a diesel engine may result in various undesired consequences. Among those, increased fuel consumption, decreased engine power, higher NO_x emission, piston ring sticking, and even cold start problems are some of the more frequently reported ones [1, 14, 18]. To tackle these disadvantages, various strategies can be applied, among which the usage of fuel additives looks to be a particularly promising one [10]. Namely, according to laboratory experiments, proper engagement of adequate fuel additives may actually result in improved engine performance and reduced exhaust emissions.

In recent years, nanomaterials (NMs) came into focus as possible novel and very promising fuel additives for diesel engines [10, 15, 17]. A diesel engine fuel with NMs additive can be obtained by dispersing NMs (particles, …) into a suitable *base fuel* and taking measures to prevent their bonding and deposition in the fuel tank or fuel pipes; for the sake of simplicity, such a fuel will be referred to as a *NMs-enriched fuel* or simply as *nanofuel* throughout this book. Furthermore, a nanofuel must exhibit uniform distribution of NMs molecules within the fuel and long-term stability of the suspension. To prepare such a stable nanofuel without NMs agglomeration, special treatment and addition of various surfactants might be necessary [3].

Till today, a respectable amount of investigations has already been done to study the effects of nanofuels usage on various diesel engine characteristics. Some researchers investigated nanofuel spray characteristics by studying spray angle, penetration, and atomization. Quite some work was related to the nanofuels combustion process, mostly based on studying in-cylinder temperature and pressure, heat release rate, and exhaust gas temperature. Most of the work, however, was related to harmful diesel engine emissions, effective power, brake-specific fuel consumption, and brake thermal efficiency [3, 6].

This book aims to provide a compact and structured overview of current knowledge and recent investigation results related to NMs additives engaged in various

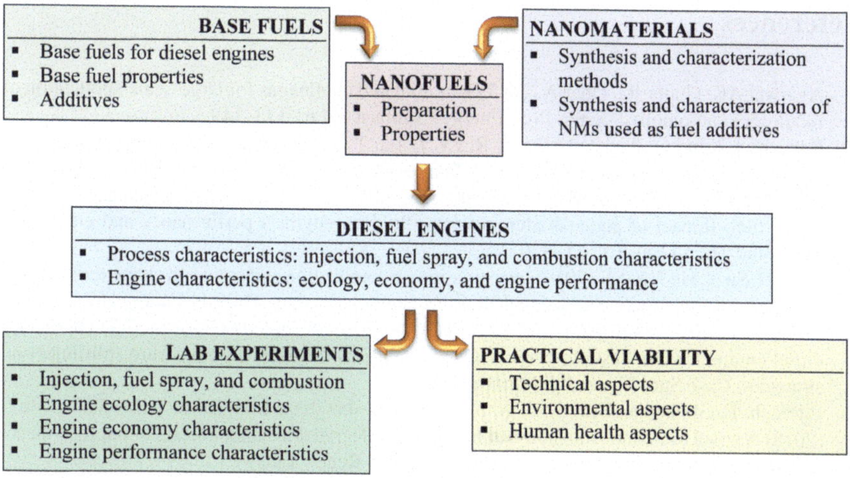

Fig. 1.1 Topics covered in this book and their interrelations

diesel engine base fuels. To provide a proper foundation, the most important diesel engine characteristics as well as NMs synthesis methods and characterization techniques are also presented. The topics covered in this book and their interrelations are illustrated in Fig. 1.1.

The structure of this book is as follows: Chap. 2 describes briefly the referred diesel engine characteristics: injection, fuel spray, and combustion characteristics, harmful diesel emissions, fuel consumption, and engine performance and presents the underlying base fuels used to prepare the nanofuels engaged in diesel engines. In Chap. 3, the synthesis methods and characterization techniques of NMs for diesel engine applications are presented; actual synthesis procedures and characterization results of NMs used as fuel additives in diesel engines are also given. Chapter 4 deals with nanofuels preparation and their physical and chemical properties; these properties are compared by those of corresponding base fuels. The book is focused on the influence of various nanofuels on the injection, fuel spray, and combustion characteristics, emission characteristics, and on engine performance. For this purpose, Chap. 5 deals with experimental laboratory results obtained with diesel engines run with NMs-enriched fuels and discusses mainly the short-term benefits of NMs fuel additives. A practical viability evaluation of nanofuels usage in a diesel engine, by considering technical, environmental, and human health aspects, is given in Chap. 6. Finally, Chap. 7 gives some conclusions and future perspectives.

References

1. Agarwal AK, Gupta JG, Dhar A (2017) Potential and challenges for large-scale application of biodiesel in automotive sector. Prog Energy Combust Sci 61:113–149
2. Baeyens J, Kang Q, Appels L, Dewil R, Lv Y, Tan T (2015) Challenges and opportunities in improving the production of bio ethanol. Prog Energy Combust Sci 47:60–88
3. Chaichan MT, Kadhum AAH, Al Amiery AA (2017) Novel technique for enhancement of diesel fuel: Impact of aqueous alumina nano-fluid on engine's performance and emissions. Case Stud Therm Eng 10:611–620
4. Chen H, Su X, He J, Xie B (2019) Investigation on combustion and emission characteristics of a common rail diesel engine fueled with diesel/n-pentanol/methanol blends. Energy 167:297–311
5. Edara G, Murthy YVVS, Srinivas P, Nayar J (2018) Effect of cooled EGR on modified light duty diesel engine for combustion, performance and emissions under high pressure split injection strategies. Case Stud Therm Eng 12:188–202
6. Ettefaghi E, Ghobadian B, Rashidi A, Najafi G, Khoshtaghaza MH, Rashtchi M, Sadeghian S (2018) A novel bio-nano emulsion fuel based on biodegradable nanoparticles to improve diesel engines performance and reduce exhaust emissions. Renew Energy 125:64–72
7. Farobie O, Matsumura Y (2017) State of the art of biodiesel production under supercritical conditions. Prog Energy Combust Sci 63:173–203
8. Geng P, Cao E, Tan Q, Wei L (2017) Effects of alternative fuels on the combustion characteristics and emission products from diesel engines: a review. Renew Sustain Energy Rev 71:523–534
9. Hoseini SS, Najafi G, Ghobadian B, Mamat R, Sidik NAC, Azmi WH (2017) The effect of combustion management on diesel engine emissions fueled with biodiesel-diesel blends. Renew Sustain Energy Rev 73:307–331
10. Hosseinzadeh-Bandbafha H, Tabatabaei M, Aghbashlo M, Khanali M, Demirbas A (2018) A comprehensive review on the environmental impacts of diesel/biodiesel additives. Energy Convers Manag 174:579–614
11. How HG, Masjuki HH, Kalam MA, Teoh YH (2018) Influence of injection timing and split injection strategy on performance, emissions, and combustion characteristics of diesel engine fueled with biodiesel blended fuels. Fuel 213:106–114
12. Kalghatgi G (2018) Is it really the end of internal combustion engines and petroleum in transport? Appl Energy 225:965–974
13. Kalghatgi G, Levinsky H, Colket M (2018) Future transportation fuels. Prog Energy Combust Sci 69:103–105
14. Kegl B, Kegl M, Pehan S (2013) Green diesel engines. Biodiesel usage in diesel engines. Springer, London
15. Khalife E, Tabatabaei M, Demirbas A, Aghbashlo M (2017) Impacts of additives on performance and emission characteristics of diesel engines during steady state operation. Prog Energy Combust Sci 59:32–78
16. Knothe G, Razon LF (2017) Biodiesel fuels. Prog Energy Combust Sci 58:36–59
17. Mehregan M, Moghiman M (2018) Effecta of nano-additives on pollutants emission and engine performance in a urea-SCR equipped diesel engine fueled with blended-biodiesel. Fuel 222:402–406
18. Soudagar MEM, Nik Ghazali NN, Kalam MA, Badruddin IA, Banapurmath NR, Akram N (2018) The effect of nano-additives in diesel-biodiesel fuel blends: a comprehensive review on stability, engine performance and emission characteristics. Energy Convers Manag 178:146–177
19. Verhelst S, Turner JWG, Sileghem L, Vancoillie J (2019) Methanol as a fuel for internal combustion engines. Prog Energy Combust Sci 70:43–88

Chapter 2
Diesel Engines

In recent years, the diesel engine sector has been one of the most active and innovative communities of technological development in the quest to improve diesel engine characteristics including ecology, economy, and engine performance. By using mineral diesel and alternative fuels, various modern technologies, including engine management and after-treatment technology, have been developed in order to improve diesel engine characteristics [68]. Most notably, precise control of the injection process, enabled by improved mechanically controlled fuel injection systems (MC-FIS) and, especially, by electronically controlled fuel injection systems (EC-FIS), enables better injection and combustion processes, and consequently delivers improved engine power, reduced fuel consumption, and reduced formation of harmful emissions.

In general, an EC-FIS comprises a high-pressure supply pump, a common rail, injectors, an ECU, and sensors (Fig. 2.1). The common rail pressure is controlled by varying fuel discharge of the high-pressure supply pump by a pump control valve. It is detected by a high-pressure sensor installed on the common rail and controlled to the predetermined value depending on engine load and speed. The common rail pressure is applied to the nozzle side of the injector as usually and also to the back side of the nozzle. The injection quantity and timing are varied by controlling the back pressure of the nozzle by means of an electromagnetically controlled solenoid valve. The injection quantity is controlled by changing the pulse width applied to the solenoid valve; the injection timing is controlled by changing the timing of the pulse. When using piezoinjectors instead of injectors with solenoid valves, fuel at high pressure is constantly delivered to the tip of the injector where a needle rests and blocks the fuel from being injected. The greatest advantage of piezoinjectors is the rate and precision in which fuel can be delivered, since the actuator can be rapidly activated and deactivated. Piezoinjectors are also known for their superior reliability. The result is increased longevity, durability, efficiency, and reduced emissions .

T. Kegl et al., *Nanomaterials for Environmental Application*, Green Energy and Technology, https://doi.org/10.1007/978-3-030-54708-0_2

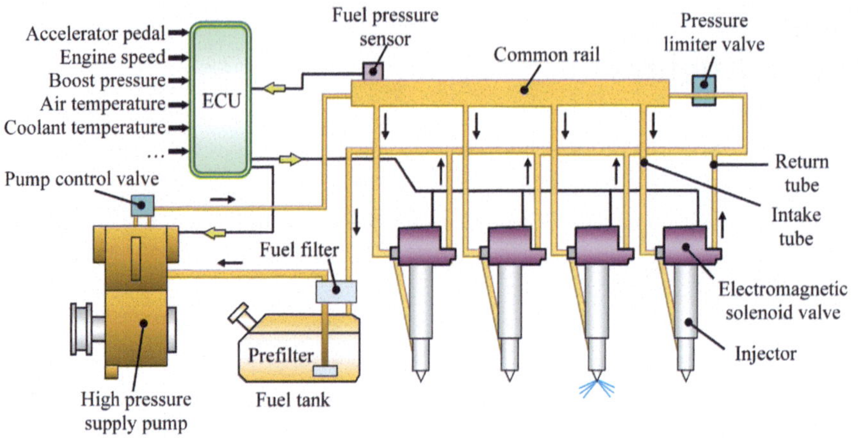

Fig. 2.1 Electronically controlled fuel injection system

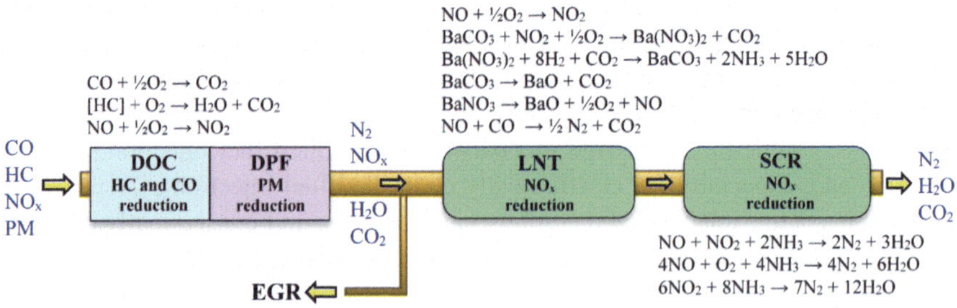

Fig. 2.2 Exhaust gas after-treatment technologies

Among various exhaust gas after-treatment technologies, the most promising are various combinations of (Fig. 2.2):

- diesel oxidation catalyst (DOC),
- lean NO_x trap (LNT),
- selective catalytic reduction (SCR),
- diesel particulate filters (DPF), and
- exhaust gas recirculation (EGR) technologies,

which may reduce the release of harmful emissions into the environment. For illustration, Fig. 2.2 shows an example assembly of after-treatment units with the corresponding chemical reactions [20, 54, 108, 143]. The reduction of HC and CO emissions in the DOC is followed by the reduction of PM emissions in the DPF and NO_x emissions reduction in the LNT and SCR units.

A modern DOC is a monolith honeycomb substrate, coated with a platinum metal group catalyst and packaged into a stainless steel container [20]. The conversion efficiency of the DOC depends on exhaust gas temperature, exhaust mass flow rate, selected catalyst, and monolith diameter and length. A DPF consists of a metallic honeycomb structure with a ceramic compound inside. Among various types of DPF, the wall flow filter seems to remove the PM in the most efficient way [20]. A LNT consists of a storage component and a precious metal supported on alumina. The LNT works under alternative cyclic conditions of lean (excess of oxygen) and rich (excess of fuel) conditions. Under the lean conditions, NO_2 is absorbed onto the catalyst and stored as a nitrate at the surface. Under fuel-rich conditions, this stored nitrate in the LNT unit is released as NH_3 and NO and then reduced to N_2. NH_3 formation by NO_x reduction within LNT causes a discontinuous NH_3 supply to the SCR. A SCR converter consists of a honeycomb monolith, coated or extruded with various catalysts ($V/W/TiO_2$, $V_2O_5/WO_3/TiO_2$, Fe-zeolites, Cu-zeolites ...) [20, 147]. The NO_x conversion efficiency of the SCR depends on the used catalyst and its configuration (coated type, extruded type), temperature, and ammonia/nitrogen ratio [145]. In order to reduce the NO_x with minimal impact on other pollutant emissions, the EGR technology is often used, which requires the determination of optimal EGR rate in dependence on current engine operating regime [135]. It may be worth noting that the presented after-treatment technologies are theoretically sufficient to completely remove CO, HC, PM, and NO_x emissions. Thus, under ideal conditions only N_2, H_2O, and CO_2 emissions would exit the exhaust system and enter the environment.

Besides of the engine management and after-treatment technologies, various alternative fuels offer another opportunity to reduce harmful emissions. This, however, is perhaps the most challenging option, due to the substantial diversity of alternative fuel properties, especially, if we take into account the wide range of possibilities offered by fuel additives.

2.1 Diesel Engine Process Characteristics

Diesel engine process characteristics are mainly related to [68]:

- injection,
- fuel spray, and
- combustion processes.

The most important injection process characteristics are injection timing (the moment of injection start, given in crankshaft angle (CA) before top dead center (TDC)), injection pressure, injection duration, and injection rate (fuel quantity injected into the combustion chamber per unit time). Fuel spray characteristics are related to spray tip penetration, spray cone angle, and droplet size. Both, injection process characteristics and fuel spray development, depend strongly on the fuel properties, type of engine and fuel injection system (FIS), and on engine operating regime. Finally, all of the mentioned characteristics and parameters, including the combustion chamber geometry, influence the combustion process characteristics such as the in-cylinder pressure, in-cylinder temperature, and heat release history.

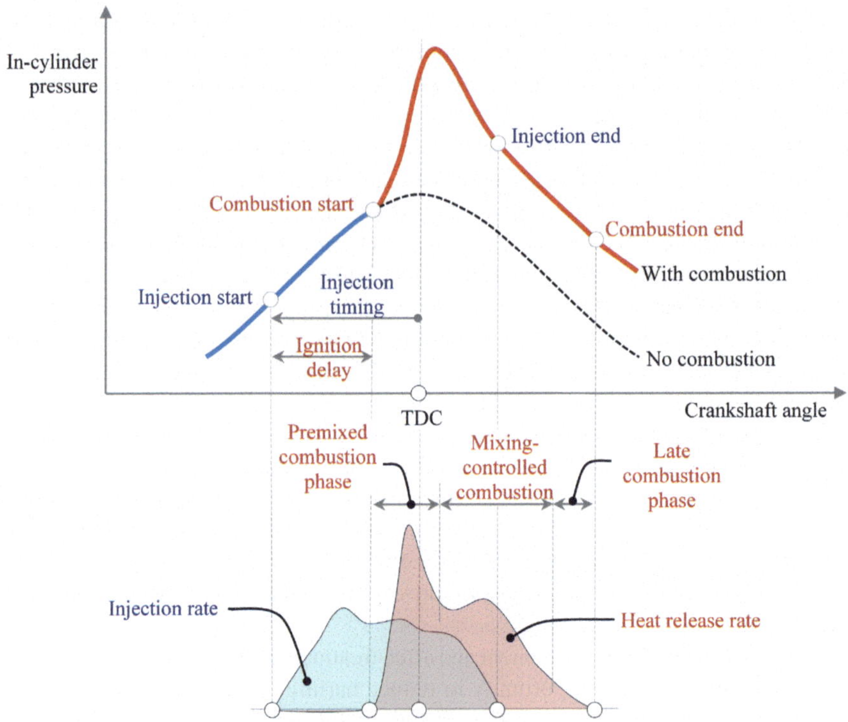

Fig. 2.3 Injection and combustion process characteristics

For illustration, typical injection rate, in-cylinder pressure, and heat release rate histories, along with related parameters, are shown in Fig. 2.3 [53, 68, 110].

Injection timing, injection pressure, and injection rate depend to a great extent on fuel properties and geometrical parameters of the fuel pump, high-pressure tube, and injector. Because fuel properties depend significantly on temperature, fuel temperature also represents a very important parameter. Fuel properties, especially the density and viscosity, influence directly the injection timing, injection pressure, and injection rate [68]. It might be worth noting that these effects are more emphasized in a MC-FIS than in an EC-FIS, such as a common rail fuel injection system (CR-FIS). Furthermore, injection timing, injection pressure, and injection rate history influence heavily the combustion process and consequently all diesel engine characteristics. For example, in a MC-FIS, where fuel injection pressure varies significantly during the injection process, the injected fuel quantity has to be as small as possible at the start of injection to reduce NO_x emission and at the end of injection interval to reduce smoke emission [63–65, 68]. On the other hand, in a CR-FIS, for example, where the fine-tuned injection pressure is practically constant through the whole injection process, precisely controlled injection timing and injection rate within a multiple injection strategy represent a very effective way to improve engine performance and reduce harmful emissions simultaneously [5, 52, 105]. In such a case, appropriate values of injection parameters can be set separately for individual engine operating regimes by the engine electronic control unit (ECU) based on the data maps stored in it [105]. These ECU data maps are obtained by engine testing and on the basis of certain known principles, often derived from MC-FIS testing. For example, it is, in generally, known that with advanced injection timing the premixed combustion increases due to increased time available for fuel evaporation, which improves the combustion. On the other way, with retarded injection timing, the majority of heat release occurs during the mixing-controlled combustion phase, which is undesirable because it reduces the overall combustion efficiency and increases the heat losses [5].

Fuel spray characteristics can be classified into microscopic (droplet size and velocity) and macroscopic (spray tip penetration, liquid spray breakup length, spray area, and spray cone angle) characteristics [59, 82]. Fuel spray characteristics are affected to a great extent by physical and chemical fuel properties, injection process, nozzle geometry, and combustion chamber conditions and geometry [67]. In general, fuels with high viscosity tend to form larger injected fuel droplets, which leads to poorer fuel atomization; this increases spray tip penetration and decreases the spray angle, which may lead to poor combustion [80]. Furthermore, higher fuel viscosity tends to inhibit cavitation inception inside the nozzle holes, which can be observed by asymmetry (inclination) of the fuel spray [67]. This also affects the spray because cavitation in the nozzle holes influences spray break-up and fuel atomization [82, 149].

In-cylinder pressure $\left(p_{\mathrm{cyl}}\right)$ and in-cylinder temperature $\left(T_{\mathrm{cyl}}\right)$ depend on the quantity of fuel burned during premixed combustion phase and on the mixing ability of fuel with air. In general, p_{cyl} and T_{cyl} increase with advanced injection timing [110]. Agarwal et al. [5] confirmed that a change in injection timing may alter the p_{cyl} and T_{cyl} at time of injection. Furthermore, it was shown that the position of maximal in-cylinder pressure $\left(p_{\mathrm{cyl,max}}\right)$ depends to a great extent on injection timing and fuel injection pressure. A higher $p_{\mathrm{cyl,max}}$ can be exploited for increasing the maximum effective pressure or improving the brake thermal efficiency of a diesel engine. Unfortunately, a higher T_{cyl} is directly related to higher NO_x emission [68].

Heat release rate ($\dot{q}_{\mathrm{h}}$) history can be divided into four main phases: ignition delay (the time span between the injection start and combustion start), premixed combustion, mixing-controlled combustion, and late combustion phase [53, 83, 103]. During the first (physical) phase of the ignition delay, the fuel is breaking up, it vaporizes, blends with air, and heats to self-ignition temperature. During the second (chemical) phase of the ignition delay, reactions begin gradually and accelerate afterward until combustion occurs. During the ignition delay period $\left(\varphi_{\mathrm{ig.d}}\right)$, a negative heat release is observed due to fuel vaporization [103]. Ignition delay determines the maximum amount of heat released during combustion in a diesel engine. Namely, as the ignition delay increases, more fuel gets mixed with air and forms a combustion-ready mixture; as the self-ignition temperature is achieved, rapid burning of this mixture takes place, resulting in a large amount of released energy [14, 15]. The ignition delay can be reduced by rising the cetane number of the fuel [32, 83, 96, 120, 142]. The start of ignition is determined by the time point when the lowest positive value of the fuel droplet temperature gradient ($\mathrm{d}T/\mathrm{d}t$) and the first successive inflection point of the fuel droplet temperature $\left(\mathrm{d}^2T/\mathrm{d}t^2 = 0\right)$ are reached [95, 132]. After combustion start, the heat release becomes positive and the maximum value of heat release rate $\left(\dot{q}_{\mathrm{h,max}}\right)$ is observed in the premixed combustion phase. The major part of combustion takes place in premixed combustion phase, and the remaining heat is released in the mixing-controlled combustion phase. Oxygen content in fuels can accelerate the combustion during the mixing-controlled combustion phase, which results in overall lower combustion duration [103]. In the late combustion phase, the heat release usually amounts to about 20% of the total fuel energy [83].

Because the combustion process depends on so many parameters, it is a rather difficult controllable process, which, however, has a decisive influence on diesel engine characteristics.

2.2 Diesel Engine Characteristics

Diesel engine characteristics are related to engine ecology, economy, and performance.

2.2.1 Ecology Characteristics

The most important diesel harmful emissions are NO_x, PM, CO, and HC emissions [6, 68]. In order to reduce these emissions, the injection timing, injection duration, injection pressure, and injection rate history have to be precisely controlled in dependence of the used fuel, operating regime, and diesel engine type [53]. Although there are some exceptions related to a specific diesel engine, for each emission some of the most influencing parameters can be identified as follows.

NO_x formation depends on the concentration of oxygen, the time available for formation, and the temperature, according to the reaction of Zeldovich's mechanism [27, 71, 120]. During the combustion period, high in-cylinder temperatures disengage the molecular bonds of nitrogen, which then takes part in a series of reactions with oxygen through the Zeldovich mechanism, resulting in thermal NO_x [120]. It follows that NO_x emission is mainly depended on in-cylinder temperature, local concentration of oxygen, and duration of combustion during different combustion phases [2, 125]. Thus, NO_x emission can be reduced by shortening the combustion duration, by lowering in-cylinder temperature (e.g., by retarded injection timing), and by early combustion start (e.g., by lowering the premixed burned fraction) [38, 68]. Advanced injection timing and a shortened premixed combustion phase can usually provide better brake thermal efficiency while constraining NO_x, especially at low-load operating regimes [120]. Furthermore, lower amount of nanofuel burned in the premixed combustion phase can reduce the in-cylinder temperature and consequently reduce the NO_x emission, which is produced mainly in the first part of the mixing-controlled combustion phase [32, 41, 45, 115].

Particulates from diesel engine exhaust include carbonaceous particulates or soot and other liquid and solid particulates [6, 144]. It is estimated that nearly 50% of diesel engine exhaust particulates are composed of soot [90], which contributes more than 90% of carbon particulates present in the atmosphere [57]. In addition to the elemental carbon in particulates' emission, the solid fraction of diesel exhaust contains metal and metal oxides, originating from engine wear and from lubrication and fuel additives [133]. Particulates from diesel engine exhaust can be categorized due to typical particulate size distribution into five sizes of the aerodynamic diameter [43]:

- large, >10 μm,
- coarse, 2.5–10 μm (PM_{10}),
- fine, 0.1–2.5 μm ($PM_{2.5}$),
- ultrafine, 50–100 nm, and
- nano-sized <50 nm.

It should be noted that the smallest particulates (within the nuclei mode) do not contribute substantially to the total particulate mass, but their count is the highest. On the other hand, the relative small number of larger particulates (within the accumulation mode) dominates the particulate mass. Beatrice et al. [23] investigated the cumulative particulate number distribution during the NEDC driving cycle, where 90% of particulates have a diameter of less than 50 nm. It turned out that these particulates represent only 1–20% of exhaust particulate mass. Approximately 60–80% of the PM mass lies in the accumulation mode, and the diameter of these particulates is in the range from 100 nm to 1 μm. The coarse mode with particulate sizes larger than 1 μm represents 5–20% of the mass, but the particulate count in this mode is minor [6, 89, 90, 92]; it might be worth noting that the smaller particulates are those considered to be the most harmful to human health [43]. Furthermore, it has to be mentioned that some particulates from diesel exhaust are visible as opaque dark-colored smoke. Lapuerta et al. [80] reported that smoke emission may be directly related to the viscosity of fuels. Namely, high viscosity fuels tend to form larger fuel droplets, which leads to poorer fuel atomization and consequently to poor combustion, resulting in increased smoke emission. In general, PM (including smoke) emission from diesel engine exhaust can be reduced by increasing oxygenate content in the fuel and by higher injection pressure, which contributes to decreased fuel droplet size and improved combustion [68, 118].

CO emission is generated mainly by exaggerated fuel richness and unavailability of oxygen in the combustion zone [35]. Besides of assuring an appropriate air/fuel ratio, there are several options for reducing CO emission efficiently. For example, injection timing influences the CO emission to a great extent. While retarded injection timing increases the CO emission due to a longer heat release phase, advanced injection timing may also increase CO emission due to a larger fuel-rich zone [53].

Furthermore, at constant injection timing, CO emission can be reduced by increasing the fuel injection pressure [53].

The unburned HC in diesel exhaust consists of original fuel molecules, products of pyrolysis of fuel compounds, and partially oxidized hydrocarbons. Several processes can contribute to unburned HC emission due to incomplete combustion. For example, proper mixing of fuel and air is very important. Namely, over-mixing of fuel and air beyond lean flammability limits during the ignition delay period can lead to higher HC emission. Furthermore, under-mixing of fuel injected toward the end of injection process results in fuel/air ratios that are too rich for complete combustion. Another reason for HC emission can also be the wall quenching of the flame due to impingement of the fuel spray on the peripheral areas of the combustion chamber [121]. Furthermore, retarded injection timing causes lower in-cylinder pressure and in-cylinder temperature and increases HC emission due to lower speed of oxidation [26, 68]. It might be worth noting that fuel properties can also effect HC emission directly. For example, fuels with high CN exhibit a shorter ignition delay, leading to a high in-cylinder temperature and increased speed of oxidation, with a consequent reduction in unburned HC emission [26].

2.2.2 *Economy Characteristics*

Economy characteristics are mainly related to brake-specific fuel consumption. Brake-specific fuel consumption is defined as $g_e = 3.6 \times 10^6 \frac{\dot{m}_f}{P_e}$, where P_e (kW) denotes effective power and $\dot{m}_f$ (kg/s) is the fuel mass flow. Brake-specific fuel consumption of a diesel engine depends on injection and combustion process characteristics and fuel properties such as the density, viscosity, and calorific value. Currently achievable values of g_e are ranged from 225 to 205 g/kWh; meanwhile, the desired value of g_e is lower than 170 g/kWh [58].

2.2.3 *Engine Performance Characteristics*

The most frequently analyzed engine performance characteristics are: effective power P_e (kW), effective torque M_e (Nm), and brake thermal efficiency η_{th} (%). Brake thermal efficiency is the ratio between the brake power output and the energy introduced through fuel injection, while the latter is the product of the injected fuel mass flow rate and the lower heating value [69, 79]. Currently achievable η_{th} is ranged from 38 to 42%; meanwhile, the desired value of η_{th} is over 50% [58]. Each of these performances is affected by fuel properties, injection and combustion design and control parameters, engine and injection type as well as by engine operating regime [36, 44, 49, 50, 53, 68, 97]. In general, η_{th} can be improved by reduced ignition delay as a consequence of higher cetane number and calorific value of fuels [31, 142, 151]. A higher cetane number results in slightly higher in-cylinder pressure and faster heat release rate, while higher calorific value promotes more complete combustion due to higher evaporation rate and higher in-cylinder temperature [31, 151].

2.3 Diesel Engine Fuels

The most frequently used fuels, serving as base fuels for preparation of diesel engine nanofuels, are:

- mineral diesel (D),
- biodiesels from various sources (*source*B),
- oils from various sources (*source*O),
- gas-to-liquid (GTL) fuels,
- blends of mineral diesel and biodiesels from various sources (D/*source*Bxx),
- blends of biodiesels from various source and oils from various sources (*source*B/*source*Oxx),
- water emulsified fuel (D/Wxx, *sourceB*/Wxx, *source*O/Wxx, D/*source*Bxx/Wyy), and
- ethanol, butanol or pentanol emulsified fuels (D/Exx, D/Buxx, D/*source*Bxx/Eyy, D/*source*Bxx/Buyy, D/*source*Bxx/Pyy, *source*B/Buxx, *source*B/Exx),

where xx and yy denote parts in % of biodiesel (*source*B), water (W), ethanol (E), butanol (Bu), or pentanol (P).

Mineral diesel (D) is produced by fractional distillation of crude oil between 200 and 350 °C at atmospheric pressure, resulting in a mixture of carbon chains that typically contain between 8 and 21 carbon atoms per molecule [22]. Perhaps the most exposed challenge in reducing pollutants from a diesel engine, fueled by mineral diesel, is related to the trade-off between NO_x and PM emissions [68].

Biodiesel (*source*B) is a renewable and environmentally friendly alternative fuel for a diesel engine. Biodiesels are sulfur-free and oxygenated fuels which contain 10–15% oxygen by weight [68]. With respect to raw materials used for their production, the most frequently used biodiesels are [3, 4, 29, 68, 73, 98]:

- *Ailanthus altissima* biodiesel (*aa*B),
- *Botryococcus braunii microalgae* biodiesel (*bba*B),
- *Calophyllum inophyllum* biodiesel (*ci*B),
- *Castor oil* biodiesel (*co*B),
- *Corn* biodiesel (*c*B),
- *Cottonseed* biodiesel (*cs*B),
- *Cashew nut shell* biodiesel (*cns*B),
- *Grape seed* biodiesel (*gs*B),
- *Honge* biodiesel (*h*B),
- *Jatropha* biodiesel (*j*B),
- *Jojoba* biodiesel (*jj*B),
- *Karanja* biodiesel (*k*B),

- *Lemongrass* biodiesel (*lg*B),
- *Linseed* biodiesel (*l*B),
- *Mahua* biodiesel (*m*B),
- *Mimusops elengi* biodiesel (*me*B),
- *Mustard oil* biodiesel (*mo*B),
- *Neem* biodiesel (*n*B),
- *Nerium oleander* biodiesel (*no*B),
- *Palm* biodiesel (*p*B),
- *Pine oil* biodiesel (*po*B),
- *Pongamia pinnata* biodiesel (*pp*B),
- *Poultry litter* biodiesel (*pl*B),
- *Rapeseed* biodiesel (*r*B),
- *Rice bran* biodiesel (*rb*B),
- *Simarouba glauca* biodiesel (*sg*B),
- *Soybean* biodiesel (*s*B),
- *Sunflower* biodiesel (*sf*B),
- *Tamarind seed* biodiesel (*ts*B),
- *Waste cooking oil* biodiesel (*wco*B),
- *Waste frying oil* biodiesel (*wfo*B), and
- *Ziziphus jujube* biodiesel (*zj*B).

The most exposed possible disadvantages of biodiesels are poor cold flow properties and oxidative stability, fuel injector coking, material compatibility (biodiesel is not compatible with some plastic materials used in pipes and seals), high price, and high NO_x emissions [4, 73, 78]. Furthermore, the effects of oxidative degradation, caused by contact with ambient air during long-term storage, raise concerns related to maintaining the quality of biodiesels [73]. Oxidative stability of biodiesels can be improved by adding natural and synthetic antioxidants at varying concentrations between 100 and 1000 ppm [10, 138]. Several natural and synthetic antioxidants, such as A-tocopherol acetate, butylated hydroxyanisole, tert-butylhydroquinone, diethylamine, and pyridoxine hydrochloride, improve the oxidative stability of biodiesels such as neem oil, canola, mango seed, soybean, and rice bran [10]. Experimental results show that various types of biodiesels have various natural levels of oxidative stability, indicating that natural antioxidants play a significant role in determining their oxidative stability [117].

Waste oils and vegetable oils (*source*O) represent interesting alternative fuels for diesel engines in public transportation and marine propulsion. Waste oil is derived from material which would otherwise need disposal. Meanwhile, vegetable oils can be produced almost everywhere in the world from various crops in relatively small plants [17, 30, 98]. For example, plastic oil (*p*O) can be produced from high-density polyethylene (HDPE) plastic waste by a pyrolysis process [134]. The characterization of this plastic oil revealed that its major constituents belong to alkenes (olefin), n-aliphatic alkanes (paraffin), cycloalkanes, and alcohols. Experimental results, obtained by Sundararajan and Bhagavathi [134], indicate that no hardware modifications are required to engage plastic oil in place of mineral diesel in a diesel engine with a mechanically controlled fuel injection system (MC-FIS). Sharma et al. [128] showed that tire pyrolysis oil (*tp*O) can be produced from truck tires by pyrolysis process. Lemongrass (*Cymbopogon flexuosus*) oil (*lg*O) can be used as a sole fuel in a diesel engine without any pretreatment processes such as pyrolysis and transesterification [13].

Gas-to-liquid (GTL) fuel has a great potential as a relatively clean alternative fuel for diesel engines. GTL derived from Fischer–Tropsch synthesis is of distinctly different characteristics than mineral diesel due to its paraffinic nature, virtually zero sulfur, high H/C ratio, and low aromatic contents [120]. GTL fuels have lower kinematic viscosity, density, and distillation temperature, while their cetane number and calorific values are higher than those of mineral diesel and various biofuels [120]. The consequences of these GTL fuel properties are good spray atomization and dispersion and good fuel evaporation, which accelerates fuel mixing with air [121]. The investigation results obtained with respect to the New European Driving Cycle (NEDC) show a potential of the GTL fuel to reduce PM and unburned HC; unfortunately, the results obtained with respect to the worldwide harmonized light duty vehicle test cycle (WLTC) indicate that PM emission may be higher than that obtained with mineral diesel [131].

Ethanol (E), butanol (Bu), and pentanol (P) emulsified diesel or biodiesel fuels have also been successfully used in diesel engines [80]. In general, ethanol can be produced chemically from petroleum or by fermentation of biomass or sugar substrates [21]. Ethanol, derived by fermentation, is a flammable, clear, and colorless liquid which can also be used as fuel [84]. Butanol is mainly produced via petrochemical processes, because the traditional industrial acetone–butanol–ethanol fermentation utilizes expensive raw materials, such as cereals, potatoes or corn [46]. Natural microbial fermentation of engineered micro-organisms and biosynthesis from glucose are the common methods employed for the production of pentanol [140]. Pentanol blends with mineral diesel and biodiesels have a relative good blend stability and show no phase separation [140]. Due to vastly different volatile components, all alcohol emulsified fuels exhibit micro-explosions during the combustion process, helping to improve atomization and fuel/air mixing [18, 19, 150].

Water (W)-emulsified fuels are generated by means of a high-energy emulsification method in the presence of surface-active agents or surfactants [139]. These high-energy methods include mechanical agitation, high-pressure homogenization, and ultrasonic and supersonic vibration. Micro-explosions of water-emulsified fuels, induced during the combustion process, help to improve atomization and mixing of the fuel with air [137].

2.3.1 Fuel Properties

The most important fuel properties, affected by NMs usage, are fuel density, kinematic viscosity, flash point, cetane number, and calorific value.

Fuel density (ρ_f) is mass per unit volume of fuel at a given temperature (measured usually at 20 or 40 °C). The density is one of the principal fuel properties, which affects the pressure wave propagation through the FIS, the fuel quantity delivered into the combustion chamber, and other important injection characteristics. Due to its relation to phenomena like cavitation, fuel density also plays an important role in design of the FIS, especially the injector nozzle [122]. Moreover, fuel density can directly influence the fuel atomization which in turn affects the brake thermal efficiency of the engine [122]. Density measurement can be carried out by using a hydrometer [2].

Kinematic viscosity (ν_f) is a measure of the internal fluid friction, which tends to resist any dynamic change in the fluid motion. Consequently, kinematic viscosity is one of the vital fuel properties due to its influence on fuel flow, fuel spray atomization, and fuel spray penetration [122]. Kinematic viscosity can be measured, for example, by using various types of viscometers [2].

Flash point (T_{fp}) temperature of fuel is the lowest temperature at which the fuel can vaporize to form and ignitable mixture in air [9]. Thus, fuels with higher flash points are safer in storage and in transit [119, 122]. For European diesel fuel, the EN 590 specification requires the flash point to be at least 55 °C. Thus, if the temperature is below 55 °C, the vapor above the liquid in a vehicle fuel tank is too lean to be flammable [9]. Biodiesel standard EN 14214 requires the flash point to be at least 101 °C. The flash point can be measured, for example, by an automated Pensky–Martens closed-cup apparatus [2].

Cetane number (CN) is a measure of the quality of fuel for a compressed ignition engine. It is defined as the percentage by volume of n-cetane in a mixture of n-cetane and α-methyl naphthalene, which has the same combustion characteristics as the tested fuel, when combustion is carried out in a standard engine under particular operating conditions. A higher CN indicates the ability of the fuel to autoignite shortly after being injected into the combustion chamber, resulting in shorter ignition delay period and smoother engine operation [2, 122]. CN can be measured on Cooperative Fuel Research (CFR) engine under prescribed standard condition.

Calorific value (H_f) of a fuel is the thermal energy released per unit quantity of fuel when the fuel is burned completely and the products of combustion are cooled back to the initial temperature of the combustible mixture. It measures the energy content in a fuel. The calorific value of a fuel can be measured, for example, in a bomb calorimeter [2].

The numerical values of the properties for various diesel engine base fuels are given in Table 2.1.

2.3.2 Fuel Additives

Perhaps the most frequently used fuel additives for diesel engines are oxygenated additives, antioxidants, cold flow improvers, lubricity improvers, cetane improvers, metal-based additives, and nanomaterials [51].

Table 2.1 Properties of various base fuels

Base fuel	$\rho_f \left(\frac{kg}{m^3}\right)$	$\nu_f \left(\frac{mm^2}{s}\right)$	T_{fp} (°C)	CN (/)	$H_f \left(\frac{MJ}{kg}\right)$	References
Mineral diesel (D)	825–855	1.3–4.1	60–80	48–57	42–45	[49, 50, 87]
Gas-to-liquid (GTL)	750–760	2.5–2.8	100–110	70–75	44–47	[59, 121]
Lemongrass oil (*lg*O)	900–910	4.5–4.7	55–60	45–50	37–38	[13]
Plastic oil (*p*O)	850–870	4.5–4.7	48–50	55–57	38–41	[134]
Tire pyrolysis oil (*tp*O)	920–930	5.4–5.5	44–48	25–30	42–44	[128]
Ailanthus altissima biodiesel (*aa*B)	870–880	4.6–4.7	160–170	48–50	–	[47, 48]
Botryococcus braunii microalgae biodiesel (*bba*B)	850–870	4.3–5.4	130–140	55–58	39–40	[55, 106, 107]
Calophyllum inophyllum biodiesel (*ci*B)	860–870	3.7–5.4	160–170	52–60	36–38	[12, 16, 37]
Castor oil biodiesel (*co*B)	880–920	6.0–10.7	160–180	55–59	37–39	[11, 91]
Corn biodiesel (*c*B)	830–850	3.3–3.5	70–80	45–50	37–39	[86]
Cottonseed biodiesel (*cs*B)	880–900	5.0–6.5	100–120	–	36–37	[104]
Cashew nut shell biodiesel (*cns*B)	840–880	4.2–4.3	130–140	51–52	37–38	[33, 109]
Grape seed biodiesel (*gs*B)	875–885	4.3–4.5	–	52–55	37–38	[25, 61, 62]
Honge biodiesel (*h*B)	–	5.6–5.8	16–180	–	36–37	[124]
Jatropha biodiesel (*j*B)	870–880	5.3–5.7	160–170	51–53	38–39	[102, 128, 130]
Jojoba biodiesel (*jj*B)	890–910	11.7–15.3	100–110	52–54	38–40	[39, 123]
Karanja biodiesel (*k*B)	890–900	8.3–9.6	110–120	54–55	35–36	[130]
Lemongrass biodiesel (*lg*B)	900–910	4.5–4.6	55–60	51–53	37–38	[35]
Linseed biodiesel (*l*B)	890–910	4.0–4.2	160–170	–	37–38	[56]
Mahua biodiesel (*m*B)	860–890	4.1–4.9	100–180	52–58	38–39	[3, 77, 114]
Mimusops elengi biodiesel (*me*B)	870–880	4.5–4.7	180–190	49–50	41–42	[75]
Mustard oil biodiesel (*mo*B)	860–870	4.2–5.8	110–130	52–54	38–39	[8, 148]

(continued)

Table 2.1 (continued)

Base fuel	$\rho_f \left(\frac{kg}{m^3}\right)$	$\nu_f \left(\frac{mm^2}{s}\right)$	T_{fp} (°C)	CN (/)	$H_f \left(\frac{MJ}{kg}\right)$	References
Neem biodiesel (*n*B)	840–950	4.2–13.5	130–180	31–56	35–38	[34, 74, 126]
Nerium oleander biodiesel (*no*B)	900–91	4.5–4.7	50–60	62–63	35–36	[35, 94]
Palm biodiesel (*p*B)	870–880	5.4–5.7	180–190	60–62	40–42	[130]
Pine oil biodiesel (*po*B)	880–890	1.8–2.1	50–70	56–58	40–41	[100]
Pongamia pinnata biodiesel (*pp*B)	800–880	5.4–5.8	170–180	53–55	37–39	[31, 129]
Poultry litter biodiesel (*pl*B)	740–780	5.3–5.5	150–160	60–62	29–35	[113]
Rapeseed biodiesel (*r*B)	870–900	4.1–5.1	120–180	51–55	40–42	[66, 99, 130]
Rice bran biodiesel (*rb*B)	870–890	6.2–6.3	130–140	51–53	37–39	[112]
Simarouba glauca biodiesel (*sg*B)	870–890	4.5–4.6	150–160	52–53	37–38	[101]
Soybean biodiesel (*s*B)	870–890	3.9–4.8	120–180	45–62	37–41	[44, 130, 136]
Sunflower biodiesel (*sf*B)	850–860	4.5–4.6	90–100	48–50	41–42	[130]
Tamarind seed biodiesel (*ts*B)	880–890	5.8–5.9	150–160	52–53	38–39	[110]
Waste cooking oil biodiesel (*wco*B)	870–890	4.6–4.8	170–180	57–63	37–39	[7, 49, 50]
Waste frying oil biodiesel (*wfo*B)	870–890	4.7–4.8	160–170	61–63	38–39	[88]
Ziziphus jujube biodiesel (*zj*B)	860–870	4.8–5.0	130–140	55–57	39–40	[1]
Ethanol (E)	780–800	1.1–1.2	10–15	5–9	26–27	[24, 40, 80]
Butanol (Bu)	810–820	2.2–2.3	25–35	15–25	33–34	[24, 40, 80]
Pentanol (P)	810–820	3.8–3.9	40–45	15–25	30–36	[86, 140]
Water (W)	896–898	0.9–1.0	–	–	–	[72]

Typical oxygenated additives are alcohols (ethanol, methanol, butanol, propanol), ethers (ethyl tert-buthyl ether, methyl tert-butyl ether, di-isopropyl ether, dimethyl ether, diethyl ether), and esters (dicarboxylic corrosive esters, ceto-acetic esters) [69]. These oxygenated additives can affect important fuel properties, e.g., viscosity, flash point, low-temperature properties (such as cloud and pour points), volatility, cetane number, as well as calorific value. The increased oxygen content of fuels enriched by oxygenated additives results in a cleaner combustion which in turn may lead to a reduction of some diesel engine emissions, especially PM and smoke

[69]. By engaging oxygenated additives, the calorific value of the fuel is usually decreased; therefore, the brake power can be reduced and specific fuel consumption can be increased essentially. Investigation results show that oxygenated additives can either increase or decrease NO_x emission. For example, these additives may increase the amount of fuel burned in the premixed combustion phase; this leads to higher in-cylinder temperatures and NO_x emission [42, 111, 146]. On the other hand, the addition of oxygenated additives can increase the enthalpy of vaporization and reduce adiabatic flame temperature of fuels, which can lead to lower in-cylinder temperature and consequently lower NO_x formation [53, 85].

Antioxidant additives, like butylated hydroxyl anisole, butylated hydroxyl toluene, ethanox, pyrogallol, diphenylamine, tert-butylhydroxyquinone, glycerol triacetate, and propyl gallate, can enhance the fuel stability by inhibiting or slowing the formation of free radicals [14, 15, 76, 93]. Antioxidant additives can increase the cetane number and flash point of the fuel and reduce its calorific value. Butylated hydroxyl anisole and butylated hydroxyl toluene can also slightly increase the brake thermal efficiency [116].

Cold flow improver additives, like polymethyl acrylate, ethylene vinyl acetate copolymer, and polymaleic anhydride, are used to improve the cold flow properties by preventing crystal growth and improving the overall crystallization behavior of the fuel [81, 141]. Cold flow improver additives can remarkably improve the low-temperature performance of biodiesel–diesel blends.

Lubricity improver additives consist of an active polar head group, forming a protective film on the surface of engine-moving metal parts, and a fuel-soluble hydrocarbon tail. They might be important to achieve normal engine operation and to prevent reduced engine lifetime. Lubricity improver additives are fatty acid esters, unsaturated fatty acids dimers, aliphatic amines, and long-chain monocarboxylic acids [51].

Cetane improver additives such as nitrites, peroxides nitrates, amyl nitrates, alkyl nitrates, di-tert-butyl peroxide, aldehydes, and tetra-azoles can increase the cetane number of a fuel and consequently result in shorter ignition delays [51]. Among the commercially used cetane improvers, alkyl nitrates, such as hexyl-nitrates, octyl-nitrates, and amyl-nitrate, are the most important ones [76].

Metal-based additives such as $CuCl_2$, $FeCl_3$, and $CoCl_2$ are combustion-improving catalyst [60, 76]. They offer some potential to reduce harmful diesel engine emissions and fuel consumption and to improve engine thermal efficiency [51, 70].

In recent years, various nanomaterials, acting as diesel engine fuel additives, gained substantial attention. According to investigation results obtained so far, these additives look to be extremely promising. This follows from the wide range of options they offer in influencing practically all important engine characteristics by affecting many processes from fuel combustion to exhaust treatment [28, 53, 127].

References

1. Aalam CS, Saravanan CG (2017) Effects of nano metal oxide blended Mahua biodiesel on CRDI diesel engine. Ain Shams Eng J 8:689–696
2. Aalam CS, Saravanan CG, Kannan M (2015) Experimental investigations on a CRDI system assisted diesel engine fueled with aluminum oxide nanoparticles blended biodiesel. Alex Eng J 54:351–358
3. Acharya N, Nanda P, Panda S, Acharya S (2017) Analysis of properties and estimation of optimum blending ratio of blended mahua biodiesel. Eng Sci Technol Int J 20:511–517
4. Agarwal AK, Gupta JG, Dhar A (2017) Potential and challenges for large–scale application of biodiesel in automotive sector. Prog Energy Combust Sci 61:113–149
5. Agarwal AK, Singh AP, Maurya RK, Shukla PC, Dhar A, Srivastava DK (2018) Combustion characteristics of a common rail direct injection engine using different fuel injection strategies. Int J Therm Sci 134:475–484
6. Agarwal AK, Dhar A, Sharma N, Shukla PC (2019) Engine exhaust particulates. Springer, Singapore
7. Aghbashlo M, Tabatabaei M, Khalife E, Najafi B, Mirsalim SM, Gharehghani A, Mohammadi P, Dadak A, Shojaei TR, Khounani Z (2017) A novel emulsion fuel containing aqueous nano cerium oxide additive in diesel–biodiesel blends to improve diesel engines performance and reduce exhaust emissions: part II—exergetic analysis. Fuel 205:262–271
8. Ahmad M, Sadia H, Zafar M, Sultana S, Khan MA, Khan Z (2012) The production and quality assessment of mustard oil biodiesel: a cultivated potential oil seed crop. Energy Sources Part A Recov Util Environ Eff 34(16):1480–1490
9. Al-Abdullah MH, Kalghatgi GT, Babiker H (2015) Flash points and volatility characteristics of gasoline/diesel blends. Fuel 153:67–69
10. Alagu K, Nagappan B, Jayaraman J, Dhas AAG (2018) Impact of antioxidant additives on the performance and emission characteristics of CI engine fuelled with B20 blend of rice bran biodiesel. Environ Sci Pollut Res 25:17634–17644
11. Amin A, Gadallah A, El-Morsi AK, El-Ibiari NN, El-Diwani GI (2016) Experimental and empirical study of diesel and castor biodiesel blending effect, on kinematic viscosity, density and calorific value. Egypt J Petrol 25(4):509–514
12. Anchupogu P, Rao LN, Banavathu B (2018) Effect of alumina nano additives into biodiesel-diesel blends on the combustion performance and emission characteristics of a diesel engine with exhaust gas recirculation. Environ Sci Pollut Res 25:23294–23306
13. Annamalai M, Dhinesh B, Nanthagopal K, Sivaramakrishnan P, Lalvani JIJ, Parthsarathy M, Annamalai K (2016) An assessment on performance, combustion and emission behavior of a diesel engine powered by ceria nanoparticle blended emulsified biofuel. Energy Convers Manag 123:372–380
14. Ashok B, Nanthagopal K, Mohan A, Johny A, Tamilarasu A (2017) Comparative analysis on the effect of zinc oxide and ethanox as additives with biodiesel in CI engine. Energy 140:352–364
15. Ashok B, Nanthagopal K, Subbarao R, Johny A, Mohan A, Tamilarasu A (2017) Experimental studies on the effect of metal oxide and antioxidant additives with *Calophyllum inophyllum* methyl ester in compression ignition engine. J Clean Prod 166:474–484
16. Ashok B, Nanthagopal K, Vignesh DS (2018) *Calophyllum inophyllum* methyl ester biodiesel blend as an alternate fuel for diesel engine applications. Alex Eng J 57:1239–1247
17. Asokan MA, Vijayan R, Prabu S (2018) A review on performance, emission and combustion characteristics of a diesel engine fueled with various vegetable oil. Indian J Chem Technol 25:225–234
18. Avulapati MM, Ganippa LC, Xia Megaritis A (2016) Puffing and micro-explosion of diesel-biodiesel-ethanol blends. Fuel 166:59–66
19. Avulapati MM, Megaritis T, Xia J, Ganippa L (2019) Experimental understanding on the dynamics of micro-explosion and puffing in ternary emulsion droplets. Fuel 239:1284–1292

20. Ayodhya AS, Narayanappa KG (2018) an overview of after-treatment systems for diesel engines. Environ Sci Pollut Res 25:35034–35047
21. Baeyens J, Kang Q, Appels L, Dewil R, Lv Y, Tan T (2015) Challenges and opportunities in improving the production of bio-ethanol. Prog Energy Combust Sci 47:60–88
22. Basu P (2018) Biomass gasification, pyrolysis and torrefaction. Practical design and theory. Elsevier, London
23. Beatrice CB, Costagliola MA, Guido C, Napolitano P, Prati MV (2017) How much regeneration events influence particle emissions of DPF-equipped vehicles? SAE international. SAE technical paper 2017-24-0144
24. Bharathiraja B, Jayamuthunagai J, Sudharsanaa T, Bhaghavi A, Praveenkumar R, Chakravarthy M, Yuvaraj D (2017) Biobutanol—an impending biofuel for future: a review on upstream and downstream processing techniques. Renew Sustain Energy Rev 68:788–807
25. Bolonio D, García-Martínez MJ, Ortega MF, Lapuerta M, Rodríguez-Fernández J, Canoira L (2019) Fatty acid ethyl esters (FAEEs) obtained from grapeseed oil: a fully renewable biofuel. Renew Energy 132:278–283
26. Cataluña R, da Silva R (2012) Effect of cetane number on specific fuel consumption and particulate matter and unburned hydrocarbon emissions from diesel engines. J Combust 738940:1–6
27. Chaichan MT, Kadhum AAH, Al-Amiery AA (2017) Novel technique for enhancement of diesel fuel: Impact of aqueous alumina nano-fluid on engine's performance and emissions. Case Stud Therm Eng 10:611–620
28. Chen AF, Adzmi MA, Adam A, Othman MF, Kamaruzzaman MK, Mrwan AG (2018) Combustion characteristics, engine performances and emissions of a diesel engine using nanoparticle-diesel fuel blends with aluminum oxide, carbon nanotubes and silicon oxide. Energy Convers Manag 171:461–477
29. Coniglio L, Bennadji H, Glaude PA, Herbinet O, Billaud F (2013) Combustion chemical kinetics of biodiesel and related compounds (methyl and ethyl esters): experiments and modeling—advances and future refinements. Prog Energy Combust Sci 39:340–382
30. Corsini A, Marchegiani A, Rispoli F, Sciulli F, Venturini P (2015) Vegetable oils as fuels in Diesel engine. Engine performance and emissions. Energy Proc 81:942–949
31. D'Silva R, Hafeez M, Fernandez J, Paloth F, Rahiz IA, Binu KG, Raju K, Thirumaleshwara B (2017) Effect of copperoxide nanoadditives on the performance and emissions characteristics of a CI engine. Energy Power 7(4):99–104
32. Dec JE (1997) A conceptual model of DI diesel combustion based on laser-sheet imaging. SAE technical paper 980873
33. Devarajan Y, Nagappan BK, Munuswamy DB (2017) Performance and emissions analysis on diesel engine fueled with cashew nut shell biodiesel and pentanol blends. Korean J Chem Eng 34(4):1021–1026
34. Devarajan Y, Munuswamy DB, Mahalingam A (2018) Influence of nano-additive on performance and emission characteristics of a diesel engine running on neat neem oil biodiesel. Environ Sci Pollut Res 25:26167–26172
35. Dhinesh B, Raj YA, Kalaiselvan C, Krishnamoorthy R (2018) A numerical and experimental assessment of a coated diesel engine powered by high-performance nano biofuel. Energy Convers Manag 171:815–824
36. Dubey A, Jain S, Gupta AK, Dole M (2018) Experimental analysis on diesel engine using Al_2O_3 nano particles with hybrid biodiesel. Int Res J Eng Technol 5(4):4095–4106
37. Elangovan T, Anbarasu G, Jeryrajkumar L (2016) Development of *Calophyllum inophyllum* biodiesel and analysis of its properties at different blends. Int J Chem Tech Res 9(4):220–229
38. El-Seesy AI, Hassan H, Ookawara S (2018) Performance, combustion, and emission characteristics of a diesel engine fueled with Jatropha methyl ester and graphene oxide additives. Energy Convers Manag 166:674–686
39. El-Seesy AI, Attia AMA, El-Battsh M (2018) The effect of Aluminum oxide nanoparticles addition with Jojoba methyl ester-diesel fuel blend on a diesel engine performance, combustion and emission characteristics. Fuel 224:147–166

40. Fayyazbakhsh A, Pirouzfar V (2017) Comprehensive overview on diesel additives to reduce emissions, enhance fuel properties and improve engine performance. Renew Sustain Energy Rev 74:891–901
41. Ferguson CR, Kirkpatrick AT (2016) Internal combustion engines. Applied thermosciences. Wiley, Chichester
42. Gill S, Tsolakis A, Herreros J, York A (2012) Diesel engine emissions improvements through the use of biodieel or oxygenated blending components. Fuel 95:578–586
43. Guan B, Zhan R, Lin H, Huang Z (2015) Review of the state-of-the-art of exhaust particulate filter technology in internal combustion engines. J Environ Manag 154:225–258
44. Heydari-Maleney K, Taghizadeh-Alisaraei A, Ghobadian B, Abbaszadeh-Mayvan A (2017) Analyzing and evaluation of carbon nanotubes additives to diesohol-B2 fuels on performance and emission of diesel engines. Fuel 196:110–123
45. Heywood JB (1988) Internal combustion engine fundamentals. McGraw-Hill, Inc., New York
46. Hijosa-Valsero M, Paniagua-Garcíaa AI, Díez-Antolínez R (2018) Industrial potato peel as a feedstock for biobutanol production. New Biotechnol 46:54–60
47. Hoseini SS, Najafi G, Ghobadian B, Mamat R, Ebadi MT, Yusaf T (2018) Novel environmentally friendly fuel: the effects of nanographene oxide additives on the performance and emission characteristics of diesel engines fuelled with *Ailanthus altissima* biodiesel. Renew Energy 125:283–294
48. Hoseini SS, Najafi G, Ghobadian B, Mamat R, Ebadi MT, Yusaf T (2018) *Ailanthus altissima* (tree of heaven) seed oil: characterisation and optimisation of ultrasonication-assisted biodiesel production. Fuel 220:621–630
49. Hosseini SH, Taghizadeh-Alisaraei A, Ghobadian B, Abbaszadeh-Mayvan A (2017) Effect of added alumina as nano-catalyst to diesel-biodiesel blends on performance and emission characteristics of CI engine. Energy 124:543–552
50. Hosseini SH, Taghizadeh-Alisaraei A, Ghobadian B, Abbaszadeh–Mayvan A (2017) Performance and emission characteristics of a CI engine fuelled with carbon nanotubes and diesel-biodiesel blends. Renew Energy 111:201–213
51. Hosseinzadeh-Bandbafha H, Tabatabaei M, Aghbashlo M, Khanali M, Demirnas A (2018) A comprehensive review on the environmental impacts of diesel/biodiesel additives. Energy Convers Manag 174:579–614
52. How HG, Masjuki HH, Kalam MA, Teoh YH (2018) Influence of injection timing and split injection strategies on performance, emissions, and combustion characteristics of diesel engine fueled with biodiesel blended fuels. Fuel 213:106–114
53. Imdadul HK, Masjuki HH, Kalam MA, Zulkifli NWM, Rashed MM, Rashedul HK, Monirul IM, Mosarof MH (2015) A comprehensive review on the assessment of fuel additive effects on combustion behavior in CI engine fuelled with diesel biodiesel blends. RSC Adv 5:67541–67567
54. Ingole AK, Dixit D, Dingare SV (2017) A review on selective catalytic reduction technique for diesel engine exhaust treatment. Int J Current Eng Technol 7:206–210
55. Islam MA, Magnusson M, Brown RJ, Ayoko GA, Nabi N, Heimann K (2013) Microalgal species selection for biodiesel production based on fuel properties derived from fatty acid profiles. Energies 6:5676–5702
56. Jayanthi P, Srinivasa RM (2016) Effect of nanoparticles additives on performance and emissions characteristics of a DI diesel engine fueled with biodiesel. Int J Adv Eng Technol 9(6):689–695
57. Jeevanandam J, Barhoum A, Chan YS, Dufresne A, Danquah MK (2018) Review on nanoparticles and nanostructured materials: history, sources, toxicity and regulations. Beilstein J Nanotechnol 9:1050–1074
58. Kalghatgi G (2018) Is it really the end of internal combustion engines and petroleum in transport? Appl Energy 225:965–974
59. Kannaiyan K, Sadr R (2017) The effects of alumina nanoparticles as fuel additives on the spray characteristics of gas-to-liquid jet fuels. Exp Thermal Fluid Sci 87:93–103

60. Kannan GR, Karvembu R, Anand R (2011) Effect of metal based additive on performance emission and combustion characteristics of diesel engine fueled with biodiesel. Appl Energy 88:3694–3709
61. Karthikeyan S, Elango A, Silaimani SM, Prathima A (2014) Role of Al_2O_3 nano additive in GSO biodiesel on the working characteristics of a CI engine. Indian J Chem Technol 21:285–289
62. Karthikeyan S, Elango A, Prathima A, Raja K (2014) Environmental effects of nano additive Co_3O_4 in Grape seed oil biofuel fueled in CI engine. Res J Chem Environ 18(5):14–18
63. Kegl B (2008) Effects of biodiesel on emissions of a bus diesel engine. Biores Technol 99:863–873
64. Kegl B (2008) Biodiesel usage at low temperature. Fuel 87:1306–1317
65. Kegl B (2011) Influence of biodiesel on engine combustion and emission characteristics. Appl Energy 88:1803–1812
66. Kegl B, Kegl M (2017) Experimental investigation review of biodiesel usage in bus diesel engine. Thermal Sci 21:639–654
67. Kegl B, Lešnik L (2018) Modeling of macroscopic mineral diesel and biodiesel spray characteristics. Fuel 222:810–820
68. Kegl B, Kegl M, Pehan S (2013) Green diesel engines. Biodiesel usage in diesel engines. Springer, London
69. Khalife E, Tabatabaei M, Demirbas A, Aghbashlo M (2017) Impacts of additives on performance and emission characteristics of diesel engines during steady state operation. Prog Energy Combust Sci 59:32–78
70. Khalife E, Tabatabei M, Najafi B, Mirsalim SM, Gharehghani A, Mohammadi P, Aghbashlo M, Ghaffari A, Khounani Z, Shojaei TR, Salleh MAM (2017) A novel emulsion fuel containing aqueous nano cerium oxide additive in diesel–biodiesel blends to improve diesel engines performance and reduce exhaust emissions: part I—experimental analysis. Fuel 207:741–750
71. Khorramshokouh S, Pirouzfar V, Kazerouni Y, Fayyazbakhsh A, Abedini R (2016) Improving the properties and engine performance of diesel-methanol-nanoparticle blend fuels via optimization of the emissions and engine performance. Energy Fuels 30:8200–8208
72. Kim D, Hong S, Jang J, Park J (2017) Determination of fluid density and viscosity by analyzing flexural wave propagations on the vibrating micro-cantilever. Sensors 17:1–10
73. Knothe G, Razon LF (2017) Biodiesel fuels. Prog Energy Combust Sci 58:36–59
74. Krishna VSR, Reddy KVK, Prasad BD (2018) Experimental investigations on LHR diesel engine using biodiesel with nano additive. Int J Latest Eng Manag Res 3(2):89–95
75. Krupakaran RL, Hariprasad T, Gopalakrishna A (2018) Impact of various blends of *Mimusops elengi* methyl esters on performance and emission characteristics of a diesel engine. Int J Green Energy 15(7):415–426
76. Kumar MV, Babu AV, Kumar PP (2018) The impacts on combustion, performance and emissions of biodiesel by using additives in direct injection diesel engine. Alex Eng J 57:509–516
77. Kumar MV, Babu AV, Kumar PR (2019) Influence of metal-based cerium oxide nanoparticle additive on performance, combustion, and emissions with biodiesel in diesel engine. Environ Sci Pollut Res 26(8):7651–7664
78. Lanjekar RD, Deshmukh D (2016) A review of the effect of the combustion of biodiesel on NO_x emission, oxidative stability and cold flow properties. Renew Sustain Energy Rev 54:1401–1441
79. Lapuerta M, Armas O, Rodríguez-Fernández J (2008) Effect of biodiesel fuels on diesel engine emissions. Prog Energy Combust Sci 34:198–223
80. Lapuerta M, Rodríguez-Fernández J, Fernández-Rodríguez D, Patiño-Camino R (2017) Modeling viscosity of butanol and ethanol blends with diesel and biodiesel fuels. Fuel 199:332–338
81. Lapuerta M, González-García I, Céspedes I, Estévez C (2019) Improvement of cold flow properties of a new biofuel derived from glycerol. Fuel 242:794–803

82. Linne M (2013) Imaging in the optically dense regions of a spray: a review of developing techniques. Prog Energy Combust Sci 39:403–440
83. Liu Y, Song C, Lv G, Cao X, Wang L, Qiao Y, Yang X (2016) Surface functional groups and sp^3/sp^2 hybridization ratios of in-cylinder soot from a diesel engine fueled with n-heptane and n-heptane/toluene. Fuel 179:108–113
84. Liu G, Zhang Q, Li H, Qureshi AS, Zhang J, Bao X, Bao J (2018) Dry biorefining maximizes the potentials of simultaneous saccharification and co-fermentation for cellulosic ethanol production. Biotechnol Bioeng 115:60–69
85. Lu X, Huang Z, Zhang W, Li D (2004) The influence of ethanol additives on the performance and combustion characteristics of diesel engines. Combust Sci Technol 176:1309–1329
86. Manigandan S, Gunasekar P, Devipriya J, Nithya S (2019) Emission and injection characteristics of corn biodiesel in diesel engine. Fuel 235:723–735
87. Mat SC, Idroas MY, Hamid MF, Zainal ZA (2018) Performance and emissions of straight vegetable oils and its blends as a fuel in diesel engine: a review. Renew Sustain Energy Rev 82:808–823
88. Mehregan M, Moghiman M (2018) Effects of nano-additives on pollutants emission and engine performance in urea-SCR equipped engine fueled with blended-biodiesel. Fuel 222:402–406
89. Mili BR, Boréave A, Meme A, Vernoux P, Leblanc M, Noël L, Raux S, D'Anna B (2018) Physico-chemical characterization of fine and ultrafine particles emitted during diesel particulate filter active regeneration of Euro 5 diesel vehicles. Environ Sci Technol 52:3312–3319
90. Mohankumar S, Senthilkumar P (2017) Particulate matter formation and its control methodologies for diesel engine: a comprehensive review. Renew Sustain Energy Rev 80:1227–1238
91. Muhammad C, Jibrin MS, Mukhtar M, Sokoto AM, Baki AS (2018) Appraisal of physico-chemical properties of castor oil biodiesel blends with petro-diesel. J Energy Environ Sustain 6:1–4
92. Myung CL, Park S (2012) Exhaust nanoparticle emissions from internal combustion engines: a review. Int J Autom Technol 13(1):9–22
93. Najafi B, Akbarian E, Lashkarpour SM, Aghbasho M, Ghaziaskar HS, Tabatabaei M (2019) Modeling of a dual fueled diesel engine operated by a novel fuel containing glycerol triacetate additive and biodiesel using artificial neural network tuned by genetic algorithm to reduce engine emissions. Energy 168:1128–1137
94. Nwakaire JN, Durugu S (2013) Determination of physio-chemical properties of oleander seed oil for biodiesel production. Nigerian J Technol 32(3):440–442
95. Ooi JB, Ismail HM, Swamy V, Wang X, Swain AK, Rajanren JR (2016) Graphite oxide nanoparticle as a diesel fuel additive for cleaner emissions and lower fuel consumption. Energy Fuels 30:1341–1353
96. Ooi JB, Ismail HM, Tan BT, Wang X (2018) Effects of graphite oxide and single-walled carbon nanotubes as diesel additives on the performance, combustion, and emission characteristics of a light-duty diesel engine. Energy 161:70–80
97. Örs I, Sarıkoç S, Atabani AE, Ünalan S, Akansu SO (2018) The effects on performance, combustion and emission characteristics of DICI engine fuelled with TiO_2 nanoparticles addition in diesel/biodiesel/n-butanol blends. Fuel 234:177–188
98. Othman MF, Adam A, Najafi G, Mamat R (2017) Green fuel as alternative fuel for diesel engine: a review. Renew Sustain Energy Rev 80:694–709
99. Özgür T, Özcanli M, Aydin K (2015) Investigation of nanoparticle additives to biodiesel for improvement of the performance and exhaust emissions in a compression ignition engine. Int J Green Energy 12:51–56
100. Panithasan MS, Gopalakichenin D, Venkadesa G, Veeraraagavan S (2019) Impact of rice husk nanoparticle on the performance and emission aspects of a diesel engine running on blends of pine oil-diesel. Environ Sci Pollut Res 26(1):282–291
101. Paramashivaiah BM, Banapurmath NR, Rajashekhar CR, Khandal SV (2018) Studies on effect of graphene nanoparticles addition in different levels with *Simarouba* biodiesel and

diesel blends on performance, combustion and emission characteristics of CI engine. Arab J Sci Eng 43:4793–4801
102. Patel HK, Kumar S (2017) Experimental analysis on performance of diesel engine using mixture of diesel and bio-diesel as a working fuel with aluminum oxide nanoparticle additive. Therm Sci Eng Progr 4:252–258
103. Patel C, Chandra K, Hwang J, Agarwal RA, Gupta N, Bae C, Gupta T, Agarwal AK (2019) Comparative compression ignition engine performance, combustion, and emission characteristics, and trace metals in particulates from Waste cooking oil, Jatropha and Karanja oil derived biodiesels. Fuel 1236:1366–1376
104. Patil DS, Chopade DA, Kumbhalkar MA (2018) Experimental investigation of effect of cerium oxide nanoparticles as a fuel additive in cottonseed biodiesel blends. MAYFEB J Mech Eng 1:1–12
105. Pham VV, Cao DT (2019) A brief review of technology solutions on fuel injection system of diesel engine to increase the power and reduce environmental pollution. J Mech Eng Res Develop 42(1):1–9
106. Piloto-Rodríguez R, Sánchez-Borroto Y, Melo-Espinosa EA (2017) Assessment of diesel engine performance when fueled with biodiesel from algae and microalgae: an overview. Renew Sustain Energy Rev 69:833–842
107. Prasad ASA, Saravanan AS, Periyasamy S, Sivakumar P (2015) Optimization of various parameters on *Botryococcus braunii* for biodiesel production using nano CaO catalyst. J Chem Pharm Sci 7:125–128
108. Praveena V, Martin MLJ (2018) A review on various after treatment techniques to reduce NO_x emissions in CI engines. J Energy Inst 91:704–720
109. Radhakrishnan S, Munuswamy DB, Devarajan Y, Arunkumar T, Mahalingam A (2018) Effect of nanoparticle on emission and performance characteristics of a diesel engine fueled with cashew nut shell biodiesel. Energy Sources Part A Recov Util Environ Eff 40(20):2485–2493
110. Raju VD, Kishore PS, Nanthagopal K, Ashok B (2018) An experimental study on the effect of nanoparticles with novel tamarind seed methyl ester for diesel engine applications. Energy Convers Manag 164:655–666
111. Rakopoulus C, Dimaratos A, Giakoumis E, Rakopoulos D (2011) Study of turbocharged diesel engine operation, pollutant emissions and combustion noise radiation during starting with bio-diesel or n-butanol diesel fuel blends. Appl Energy 88:3905–3916
112. Ramanan MV, Yuvarajan D (2016) Emission analysis on the influence of magnetite nanofluid on methyl ester in diesel engine. Atmosp Pollut Res 7:477–481
113. Ramesh DK, Dhananjaya KJL, Hemanth KSG, Namith V, Parashuram BJ, Sharath S (2018) Study on effects of alumina nanoparticles as additive with poultry litter biodiesel on performance, combustion and emission characteristic of diesel engine. Mater Today Proc 5:1114–1120
114. Rani GJ, Rao YVH, Balakrishna B (2018) Investigation of fuel injection pressure impact on CI engine performance and emissions using biodiesel blend with alumina nano additives. Int J Mech Eng Technol 9(6):922–928
115. Ren Y, Huang Z, Miao H, Jiang D, Zeng K, Liu B, Wang X (2007) Effect of the addition of diglyme in diesel fuel on combustion and emissions in a compression-ignition engine. Energy Fuels 21:2573–2583
116. Rizwanul Fattah IM, Masjuki HH, Kalam MA, Mofijur M, Abedin MJ (2014) Effect of antioxidant on the performance and emission characteristics of a diesel engine fueled with palm biodiesel blends. Energy Convers Manag 79:265–272
117. Romagnoli ÉS, Borsato D, Silva LRC, Chendynski LT, Angilelli KG, Canesin EA (2018) Kinetic parameters of the oxidation reaction of commercial biodiesel with natural antioxidant additives. Ind Crops Prod 125:59–64
118. Sajeevan AC, Sajith V (2013) Diesel engine emission reduction using catalytic nanoparticles: an experimental investigation. J Eng 589382:1–9
119. Sajith V, Sobhan CB, Peterson GP (2010) Experimental investigations on the effects of cerium oxide nanoparticle fuel additives on biodiesel. Adv Mech Eng 581407:1–6

120. Sajjad H, Masjuki HH, Varman M, Kalam MA, Arbab MI, Imtenan S, Rahman SMA (2014) Engine combustion, performance and emission characteristics of gas to liquid (GTL) fuels and its blends with diesel and bio-diesel. Renew Sustain Energy Rev 30:961–986
121. Sajjad H, Masjuki HH, Varman M, Kalam MA, Arbab MI, Imtenan S, Rashed MM (2014) Comparative study of gas-to-liquid fuel, B5 diesel and their blends with respect to fuel properties, engine performance and exhaust emissions. ACS Adv 4:44529–44536
122. Sakthivel R, Ramesh K, Purnachandran R, Shameer PM (2018) A review on the properties, performance and emission aspects of the third generation biodiesels. Renew Sustain Energy Rev 82:2970–2992
123. Saleh HE, Selim MYE (2017) Improving the performance and emission characteristics of a diesel engine fueled by jojoba methyl ester-diesel-ethanol ternary blends. Fuel 207:690–701
124. Sandeep K, Rajashekhar CR, Karthik SR (2018) Experimental studies on effect of nano particle blended biodiesel combustion on performance and emission of CI engine. IOP Conf Ser Mater Sci Eng 376:012019
125. Saraee HS, Jafarmadar S, Taghavifar H, Ashrafi SJ (2015) Reduction of emissions and fuel consumption in a compression ignition engine using nanoparticles. Int J Environ Sci Technol 12:2245–2252
126. Sathiyamoorthi R, Puviyarasan M, Kumar BB, Joshua DB (2016) Effect of CeO_2 nano additive on performance and emission characteristics of diesel engine fuelled by neem oil-biodiesel. Int J Chem Sci 14(S2):473–484
127. Shaafi T, Sairam K, Gopinath A, Kumaresan G, Velraj R (2015) Effect of dispersion of various nanoadditives on the performance and emission characteristics of a CI engine fueled with diesel, biodiesel and blends—a review. Renew Sustain Energy Rev 49:563–573
128. Sharma SKS, Das RK, Sharma A (2016) Improvement in the performance and emission characteristics of diesel engine fueled with jatropha methyl ester and tyre pyrolysis oil by addition of nano additives. J Brazil Soc Mech Sci Eng 38:1907–1920
129. Sivakumar M, Sundaram NS, Kumar RR, Thasthagir MHS (2018) Effect of aluminium oxide nanoparticles blended pongamia methyl ester on performance, combustion and emission characteristics of diesel engine. Renew Energy 116:518–526
130. Sivaramakrishnan K, Ravikumar P (2012) Determination of cetane number of biodiesel and it's influence on physical properties. ARPN J Eng Appl Sci 7(2):205–211
131. Soriano JA, García-Contreras R, Leiva-Candia D, Soto F (2018) Influence on performance and emissions of an automotive diesel engine fueled with biodiesel and paraffinic fuels: GTL and biojet fuel farnesane. Energy Fuels 32:5125–5133
132. Soudagar MEM, Nik Ghazali NN, Kalam MdA, Badruddin IA, Banapurmath NR, Akram N (2018) The effect of nano-additives in diesel-biodiesel fuel blends: a comprehensive review on stability, engine performance and emission characteristics. Energy Convers Manag 178:146–177
133. Steiner S, Bisig C, Petri-Fink A, Rothen-Rutishauser B (2016) Diesel exhaust: current knowledge of adverse effects and underlying cellular mechanisms. Arch Toxicol 90:1541–1553
134. Sundararajan NK, Bhagavathi AR (2016) Experimental investigation on thermocatalytic pyrolysis of HDPE plastic waste and the effects of its liquid yield over the performance, emission, and combustion characteristics of ci engine. Energy Fuels 30(7):5379–5390
135. Thangaraja J, Kannan C (2016) Effect of exhaust gas recirculation on advanced diesel combustion and alternate fuels—a review. Appl Energy 180:169–184
136. Vellaiyan S (2019) Enhancement in combustion, performance, and emission characteristics of a diesel engine fueled with diesel, biodiesel, and its blends by using nanoadditive. Environ Sci Pollut Res 26(10):9561–9573
137. Vellaiyan S, Amirthagadeswaran KS (2016) The role of water-in-diesel emulsion and its additives on diesel engine performance and emission levels: a retrospective review. Alex Eng J 55:2463–2472
138. Velmurugan K, Sathiyagnanam AP (2016) Impact of antioxidants on NO_x emissions from a mango seed biodiesel powered DI diesel engine. Alex Eng J 55:715–722

139. Vigneswaran R, Annamalai K, Dhinesh B, Krishnamoorthy R (2018) Experimental investigation of unmodified diesel engine performance, combustion and emission with multipurpose additive along with water-in-diesel emulsion fuel. Energy Convers Manag 172:370–380
140. Vinod BM, Madhu MK, Amba PRG (2017) Butanol and pentanol: the promising biofuels for CI engines—a review. Renew Sustain Energy Rev 78:1068–1088
141. Wang J, Cao L, Han S (2014) Effect of polymeric cold flow improvers on flow properties of biodiesel from waste cooking oil. Fuel 117:876–881
142. Westbrook CK (2013) Biofuels combustion. Annu Rev Phys Chem 64:201–219
143. Wittka T, Holderbaum B, Dittman P (2015) Experimental investigation of combined LNT + SCR diesel exhaust aftertreatment. Emiss Control Sci Technol 1:167–182
144. Wu Q, Xie X, Wang Y, Roskilly T (2018) Effect of carbon coated aluminum nanoparticles as additive to biodiesel-diesel blends on performance and emission characteristics of diesel engine. Appl Energy 221:597–604
145. Xu HT, Luo ZQ, Wang N, Qu ZG, Chen J, An L (2019) Experimental study of the selective catalytic reduction after-treatment for the exhaust emission of a diesel engine. Appl Therm Eng 147:198–204
146. Yilmaz N (2012) Comparative analysis of biodiesel-ethanol-diesel and biodiesel-methanol-diesel blends in a diesel engine. Energy 40:210–213
147. Yuan X, Liu H, Gao Y (2015) Diesel engine SCR control: current development and future challenges. Emiss Control Sci Technol 1:121–133
148. Yuvarajan D, Babu MD, Kumar NB, Kishore PA (2018) Experimental investigation on the influence of titanium dioxide nanofluid on emission pattern of biodiesel in a diesel engine. Atmosp Pollut Res 9:47–52
149. Zhang X, He Z, Wang Q, Tao X, Zhou Z, Xia X, Zhang W (2018) Effect of fuel temperature on cavitation flow inside vertical multi-hole nozzles and spray characteristics with different nozzle geometries. Exp Therm Fluid Sci 91:374–387
150. Zhang Y, Huang R, Huang Y, Huang S, Zhou P, Chen X, Qin T (2018) Experimental study on combustion characteristics of an n-butanol-biodiesel droplet. Energy 160:490–499
151. Zhu M, Ma Y, Zhang D (2012) Effect of a homogeneous combustion catalyst on the combustion characteristics and fuel efficiency in a diesel engine. Appl Energy 91:166–172

Chapter 3
Nanomaterials for Diesel Engine Applications

Physicochemical properties of NMs differ to those of corresponding bulk materials. These differences are a consequence of quantum confinement of electrons within small particles and a high ratio of surface and near surface atoms or ions to the ones localized inside the particle. For this reason, NMs have a large specific surface area, resulting in an increased number of unoccupied coordination sites, defects, and crystal lattice strains [82]. Furthermore, surface atoms and ions are situated in different coordination environment, which leads to different physicochemical properties, such as spectroscopic properties, melting point, density, solubility, mechanical properties (friction, ductility, elasticity), electrical conductivity, surface tension, response to the applied magnetic field, crystal structure (some structures are stable only in nanoscale), and catalytic activity [82]. Due to these exciting facts, NMs can often be more useful than their bulk equivalents [63, 84, 85].

Engineered NMs come in the following geometrical forms:

- nanoparticles (NPs),
- nanotubes (NTs),
- nanofibers (NFs), or
- nanoplatelets (NPTs).

Throughout this paper, all of these forms will be referred to as *nanoparticles* if convenient for the sake of brevity. These NMs can be synthesized either by a "top-down" or a "bottom-up" approach [55, 58, 66]. In a top-down approach, NMs are produced by size reduction from suitable starting materials, by using various physical methods like, for example, the arc discharge method, pulse laser ablation, evaporation-condensation, spray pyrolysis, vapor-phase synthesis, pulse wire discharge, lithography, and mechanical/ball milling process [58, 66, 74]. In a bottom-up approach, NMs are built from smaller entities, by using various chemical and so-called green methods. Chemical methods are, for example, sol-gel, sol-gel combustion, hydrothermal, solvothermal, sonochemical, micro-emulsion, electrochemical, pyrolysis, microwave, photochemical, and coprecipitation methods [23, 36, 64, 66,

T. Kegl et al., *Nanomaterials for Environmental Application*, Green Energy and Technology, https://doi.org/10.1007/978-3-030-54708-0_3

91]. Green methods, which are environmentally friendly, cost effective, and easily scaled up for vast-scale synthesis, can be categorized as:

- biosynthesis methods using micro-organisms (fungi, yeasts, bacteria),
- phytosynthesis methods, using plants and plants extracts (leaves, resins, root, flowers, fruits),
- green methods which can use various templates (membranes, viruses, DNA, and diatoms) [58, 74], and
- template-free green methods [1].

3.1 Synthesis Methods

The most frequently used methods for the synthesis of NMs, potentially usable as diesel engine fuel additives, are assembled in Fig. 3.1; these methods are briefly described in the following sections.

3.1.1 Sol-Gel Method

The sol-gel method is a well-established colloidal chemistry technology, which offers a possibility to produce various NMs with novel predefined properties in a simple process at relatively low cost. This method enables preparation of highly pure NMs using inorganic compounds or metallic alkoxide sources as precursors as shown in Fig. 3.2. The sol-gel processing steps are: hydrolysis, condensation, aging, and drying [60]. After hydrolysis, partial condensation, and poly-condensation of precursor solution, the colloid (sol-gel solution) with metal-oxo-metal (M–O–M) or metal-hydroxy-metal (M–OH–M) bonds is formed. Aging and drying of NMs are carried out by evaporation, heat treatment, and sometimes by calcination at temperatures up to 800 °C [20]. The main benefits of sol-gel processing are the high purity and uniform nanostructure achievable at low temperatures.

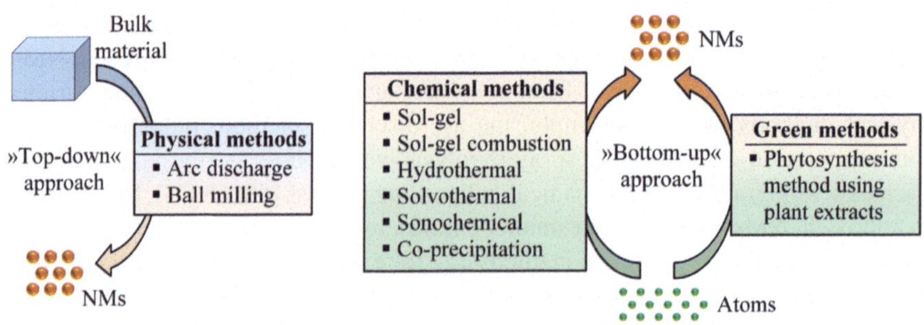

Fig. 3.1 Synthesis methods for NMs usable as fuel additives in a diesel engine

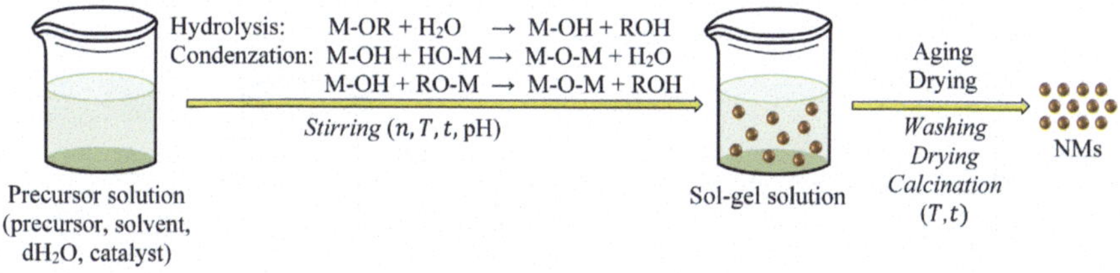

Fig. 3.2 Sol-gel method

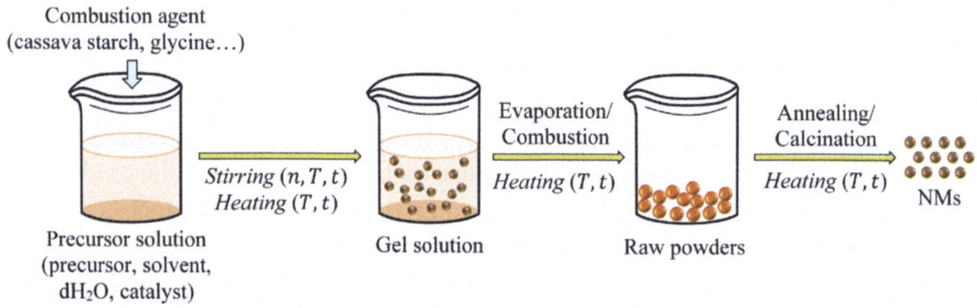

Fig. 3.3 Sol-gel combustion method

3.1.2 Sol-Gel Combustion Method

The sol-gel combustion method offers short processing times to obtain a highly crystalline material with nanoscale size and large surface area [3, 61, 80]. It is based on gelling and subsequent combustion of an aqueous solution containing salts and a combustion agent, such as citric acid, oxalic acid, cassava starch, glycine, or urea. All reactants must be well dispersed to assure a homogeneous reaction mixture. Because the reactants are in a much higher reactive state than in the corresponding solid-state reaction precursors, the reaction requires less energy and lower initial reaction threshold temperature [46]. The combustion agent forms complexes with metallic ions preventing the precipitation of hydroxilated compounds. Rapid heating of the solution containing redox groups causes a thermally induced redox reaction. During combustion, the temperature reaches approximately 400–650 °C for a few minutes [78]. Thus, the energy from the exothermic reaction between oxidant and reductant is high enough to form fine NMs. The sol-gel combustion technique is an extremely facile, time saving, and energy-efficient method, Fig. 3.3 [107].

3.1.3 Hydrothermal and Solvothermal Methods

The hydrothermal and solvothermal methods are the most common and effective synthesis methods to produce NMs with a variety of morphologies, Fig. 3.4 [41, 51,

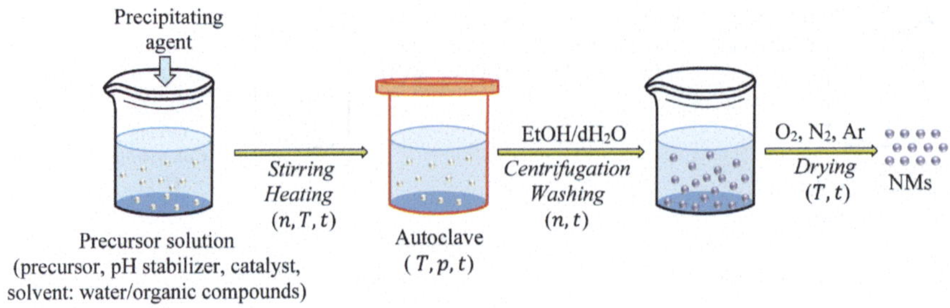

Fig. 3.4 Hydrothermal and solvothermal methods

52]. By the hydrothermal method, the reaction is carried out at high temperature and high pressure [37, 51, 52]. This process involves hydrolysis of a metallic salt and condensation of a metallic hydroxide to produce metallic NMs or metallic oxide NMs [59]. By using superheated or supercritical water, the hydrothermal process enables controlled nucleation to make narrow size NMs of a specific crystal structure. In this process, all reagents such as metallic salt solution (precursor), superheated water as solvent, pH value stabilizer, and precipitating agent, are mixed together and introduced into an autoclave, where a hydrothermal reaction is carried out [16, 21]; the reaction time and temperature are important parameters in determining the final NMs morphology [21]. The hydrothermal method is a simple method for synthetization of a wide diversity of organized structures, which are difficult to obtain with other available techniques [52]. The solvothermal method is very similar to the hydrothermal method except that the primary solvent used is not water (although water is sometimes also added to induce hydrolysis). Namely, the choice of the solvent has been found to dramatically influence the shape and size of the formed crystallites [2, 17]. The solvothermal method includes heterogeneous chemical reaction in an autoclave in presence of an organic compound, acting as a solvent at temperature above the boiling temperature of the solvent used; the processing time, initial pH value, pH value-adjusting agents, surfactant, and precursor are important parameters determining the final NMs structure and morphology [2]. The main advantages of the solvothermal synthesis are related to the simple and precise control of the size, shape distribution, and crystallinity of NMs [73]. The main disadvantages are related to cost of expensive autoclaves, safety issues during the reaction process, and to inability to observe the reaction process [10].

3.1.4 Sonochemical Method

The sonochemical method uses ultrasound for the production of NMs without high temperatures, high pressures, or long reaction times as shown in Fig. 3.5. The mechanism causing the sonochemical effects in liquids is the phenomenon of acoustic cavitation. Acoustic cavitation can locally create a temperature of around 5000 °C

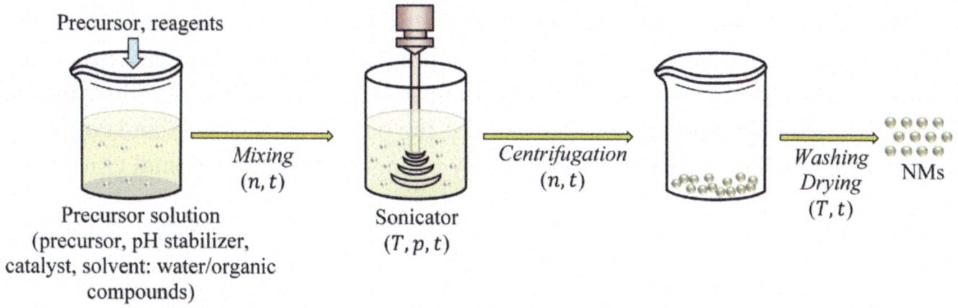

Fig. 3.5 Sonochemical method

and a pressure of more than 18 bar, which facilitates many unusual chemical reactions [4]. Most notable consequences of acoustic cavitation are related to formation, growth, and implosive collapse of bubbles. The related processes can be categorized as primary sonochemistry (gas-phase chemistry occurring inside collapsing bubbles), secondary sonochemistry (solution-phase chemistry occurring outside the bubbles), and physical modifications (caused by high-speed jets or shock waves derived from bubble collapse) [34, 76, 102]. The main advantages of the sonochemical method are its simplicity, effectiveness, mildness, and high selectiveness. Additionally, it is surfactant-free and so it does not require any additional organic solvents, water, or templating agents. Moreover, it also allows for an effective shape control of the product [4, 26].

3.1.5 Coprecipitation Method

The coprecipitation method is a widely used method for the synthesis of metallic oxide NMs from various salt solutions (chlorides, sulfates, nitrates …) and alkaline solutions (NH_4OH, NaOH) by reactions which involve simultaneous occurrence of nucleation, growth, coarsening, and/or agglomeration processes, Fig. 3.6 [42–44, 101]. After the addition of alkaline to salt solution under vigorous stirring, precipitates are obtained which are usually repeatedly washed with dH_2O and EtOH, dried,

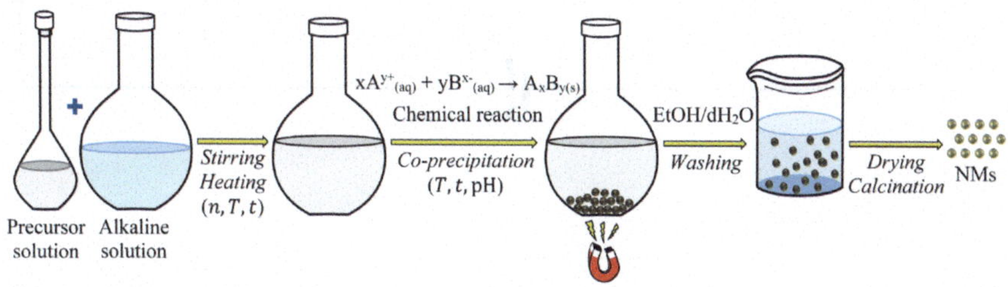

Fig. 3.6 Coprecipitation method

and sometimes calcined [69]. Advantages of this method are its simplicity, good efficiency, low chemical hazard, low reaction temperature, and short reaction time. Meanwhile, the disadvantages are related to insufficient control of size distribution and shape of generated particles [47]. Namely, the particle size, shape, and composition can be controlled only to some extent by changing the experimental conditions [101].

3.1.6 Arc Discharge Method

The arc discharge method is suitable for the synthesis of metallic, metallic oxide, and single-wall and multi-wall carbon NMs [8, 83, 85, 87]. It is based on a chamber equipped with two high-voltage electrodes, which are kept in proximity to each other in a suitable solution as shown in Fig. 3.7. The diameter of the cathode is always larger than the one of the anode, which is filled with an adequate precursor and catalyst. The chamber is filled with gas or submerged inside a liquid environment [8]. Due to the high potential difference and proximity, an electrical breakdown occurs in the medium, giving rise to arc discharge which produces an ongoing thermal plasma discharge. The local temperature of the ongoing plasma can reach several 1000 °C, which effectively vaporizes the electrode surfaces. The vaporized material condenses on the solvent to form NMs. The condensing solvent medium, voltage and current supplied to the electrode, and electrode material are the major parameters in determining the end product. The extremely high temperature produced between the electrodes can also facilitate atomic assembly pathways that can create new nanostructures which differ from the electrode materials. The main advantages of the arc discharge method are its simplicity, low-cost, and absence of multi-step purification. Meanwhile, the very low yield is the main disadvantage [83, 87].

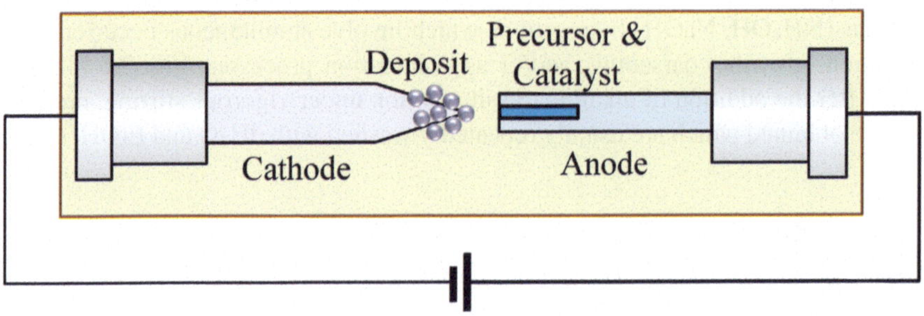

Fig. 3.7 Arc discharge method

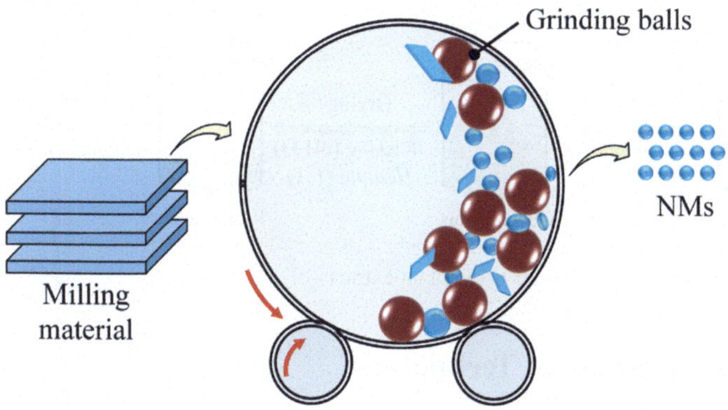

Fig. 3.8 Mechanical ball milling process

3.1.7 Mechanical Ball Milling Process

The mechanical ball milling process can be used for the synthesis of various NMs of various shapes and sizes [22, 96, 97]. The central process is the ball–powder–ball collision. Powder particles are trapped between the colliding balls during milling and undergo deformation and/or fracture processes which define the ultimate structure of the powder, Fig. 3.8. The nature of these processes depends upon the mechanical behavior of the powder components, their phase equilibria, and various milling parameters [103]. The most important milling parameters that affect the physical and morphological properties of the resulting NMs, are milling media, milling speed, ball-to-powder weight ratio, type of milling (wet or dry), type of ball mill (vibrator mill, planetary mill, attritor mill), milling atmosphere, and duration of milling [22, 53]. The mechanical ball milling method is an inexpensive and simple method, and it is free of any solvent contamination [22].

3.1.8 Green Synthesis Method Using Plant Extracts

The green synthesis method using plant extracts can be used for the synthesis of metallic NMs [84, 88, 93]. After preparation of plant extracts, a metallic salt solution with the plant extract is stirred magnetically at some prescribed temperature for a prescribed time period. The complex formed can then be centrifuged, washed, dried, and sometimes calcinated in a muffle furnace as shown in Fig. 3.9. The green synthesis method can be used for the synthesis of highly stable NMs with controlled toxicity, where NMs sizes and morphologies are controlled by altering conditions such as pH-value, substrate concentration, temperature, mixing speed, and exposure time [74].

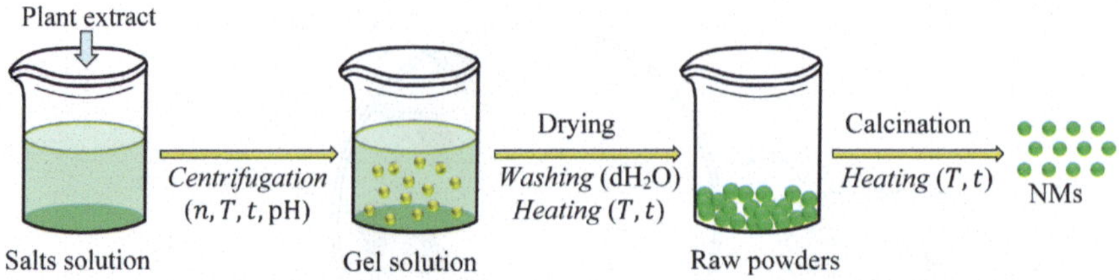

Fig. 3.9 Green synthesis methods using plant extracts

3.2 Characterization Techniques

Characterization of NMs is a process used to determine the successfulness of the NMs synthesis and to analyze the structure and the morphological, thermogravimetric, and magnetic properties of NMs. The most frequently used characterization methods for NMs, potentially usable in diesel engine fuels, are the scanning electron microscopy (SEM), transmission electron microscopy (TEM), energy-dispersive X-ray spectroscopy (EDXS), X-ray diffraction (XRD), thermogravimetric analysis (TGA), Fourier transform infrared (FT-IR) spectroscopy, and vibrating sample magnetometry (VSM). The measured parameter, principle, and some advantages and disadvantages of these techniques are given in Table 3.1 [47, 56].

3.2.1 SEM and TEM Techniques

The scanning electron microscopy (SEM) and transmission electron microscopy (TEM) are the most common methods used in morphology evaluation, Fig. 3.10 [35]. SEM and TEM are used for analyzing nanoparticle properties such as their size, shape, and texture [13, 42–44]. The resolution of SEM is around 1–2 nm, meanwhile the TEM resolution is around 0.1–0.2 nm. Whereas SEM allows investigation of the surface appearance, TEM allows examination of the internal structure of the material. TEM can be used to determine the core@shell nanostructure, layering in the structure, particle size, and surface roughness [13, 15]. In SEM, an electron beam scans the surface of the sample, which can be of any thickness and is mounted on a suitable stub. The specimen stage is at the bottom of the column, and the obtained images are shown on a monitor [35]. In TEM, an electron beam passes through a thin sample, which has to be specially prepared. The specimen stage is halfway down the column, and the images are shown on a screen [15].

Table 3.1 NMs characterization techniques

Technique	Measured parameter	Principle	Advantages (+) and disadvantages (−)
SEM	Morphology and topography	Interaction of electrons with matter	+ Direct observation of the NMs − Time consuming, require large number of NMs − Not a good technique for characterization of core@shell NPs because this technique reports total particle size
TEM	Morphology, crystallography structure, defects	Interaction of electrons with matter	+ Efficient and versatile + Good technique for characterization of core@shell NPs − Time consuming, require large number of NMs
EDXS	Elemental composition	Electromagnetic emission spectrum	+ Fast technique
XRD	Crystalline phases, crystal size	Incident rays with the sample produces constructive interference	+ Possibility to calculate the mean particle size + Non-destructive technique − X-rays do not interact very strongly with lighter elements
TGA	Mass	Weight loss during heating or cooling due to oxidation, decomposition…	+ Quantitative technique − Cannot identify decomposition products
FT-IR	Molecular bonds and functional group of molecule	Absorption of IR light by the bonds of a vibrating molecule	+ Fingerprint information of NMs
VSM	Magnetism	The induced voltage due to vibration is proportional to the sample's magnetic moment	+ Consider the magnetic history + Record the hysteresis curve − Time consuming

3.2.2 EDXS Technique

The energy-dispersive X-ray spectroscopy (EDXS) is used to provide the elemental analysis and to determine the chemical composition of the NMs [71]. Its characterization capabilities are based on the fundamental principle that each element has a unique atomic structure resulting in a unique set of peaks in its electromagnetic emission spectrum. Figure 3.11 illustrates the basic principle of operation of an EDXS [15, 71]. At first, an electron beam hits the sample and transfers part of its energy to the atoms of the sample. This energy can be used by the electrons of the atoms

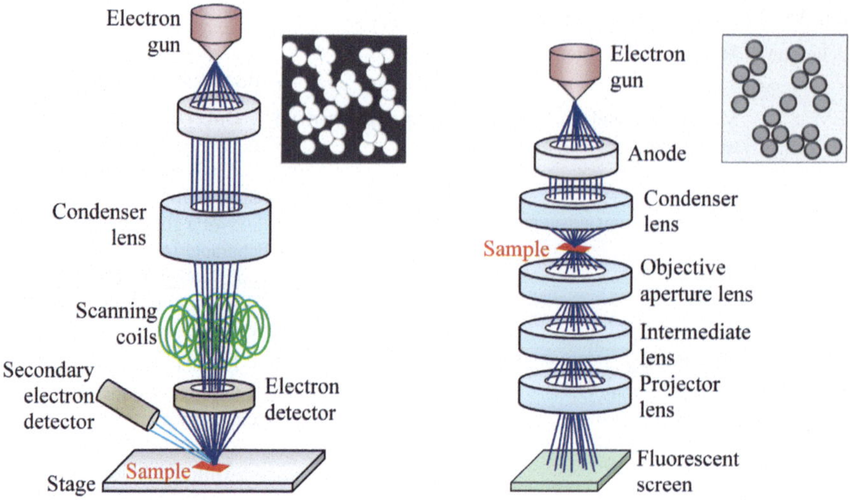

Fig. 3.10 SEM and TEM techniques

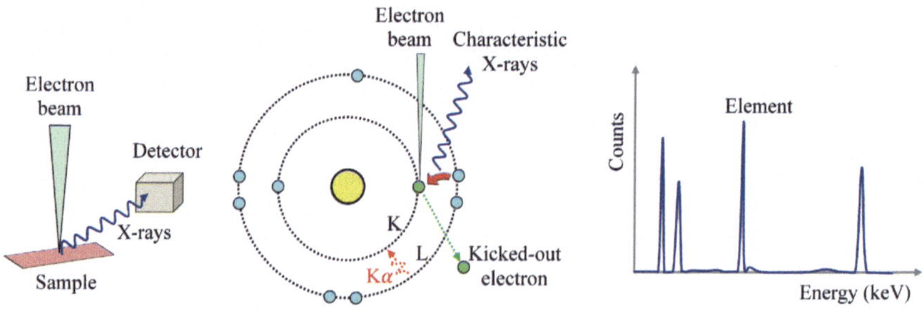

Fig. 3.11 EDXS technique

to "jump" to an energy shell with higher energy or be knocked-off from the atom. If such a transition occurs, the electron leaves behind a hole, which has a positive charge. This hole attracts the negatively charged electrons from higher-energy shells. When an electron from such a higher-energy shell fills the hole of the lower-energy shell, the energy difference of this transition is released in the form of an X-ray. The X-ray is emitted from the sample with the energy depending on the material of the sample (NMs); these X-rays are detected by a suitable system.

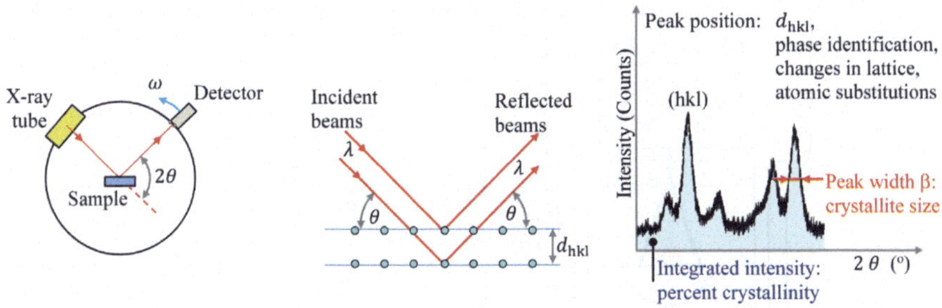

Fig. 3.12 XRD technique

3.2.3 *XRD Technique*

The X-ray diffraction (XRD) is a technique able to deliver information regarding the crystalline nature of a material, lattice parameters, and crystallite size, Fig. 3.12 [15, 35]. In this method, X-rays are generated by a cathode ray tube, filtered to produce monochromatic radiation, collimated to concentrate, and directed toward the sample. The detector is moving in a circle around the sample, and its position is recorded as the angle 2θ. The X-ray intensity is usually recorded in a diffractogram as a function of the 2θ angle as shown in Fig. 3.12. The interaction of X-ray radiation with crystalline sample is governed by Bragg's law: $n_B\lambda = 2d_{hkl}\sin\theta$, where n_B is an integer, λ is the X-ray wavelength, θ denotes the angle between the incident beams and the scattering planes, and d_{hkl} is the interplanar spacing of the crystal as shown in Fig. 3.12 [11, 35]. The Miller indices (*hkl*) of the diffraction peaks from the diffractogram are determined from the published reference pattern. On the basis of (*hkl*) and d_{hkl}, it is possible to calculate the unit cell dimensions. Besides the crystalline structure, determined on the base of the peaks intensities, the crystallite size can be determined on the basis of the peak widths. The average crystallite size d_{NPs} can be calculated at the most exposed peaks by using the Debye–Sherrer equation: $d_{NPs} = \frac{k\cdot\lambda}{\beta\cdot\cos\theta}$, where k is the Sherrer constant and β is full peak width at half of the maximum intensity [14, 35, 42–44]. On the basis of the integrated intensity in the diffractogram (Fig. 3.12), the percentage of crystallinity can be determined [14].

3.2.4 *TGA Technique*

The thermogravimetric analysis (TGA) technique can be used to determine the structure and/or composition of NMs by measuring the temperature-dependent change of the nanomaterial mass: either mass loss due to decomposition, evaporation, reduction, and desorption, or, mass gain due to oxidation and absorption Fig. 3.13 [13, 42–44]. TGA measures the amount and rate of NMs mass change with respect to temperature and time in a controlled environment. It can be used to determine the thermal and

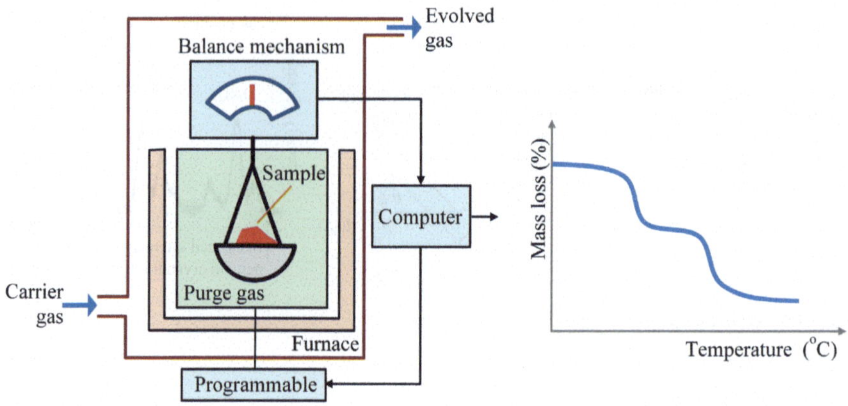

Fig. 3.13 TGA technique

oxidative stability of materials, composition of multi-component systems, estimated product lifetime, decomposition kinetics of materials, effects of reactive or corrosive atmospheres on the material, and moisture and volatiles content of materials [15, 98].

3.2.5 *FT-IR Spectroscopy*

The Fourier transform infrared (FT-IR) spectroscopy offers quantitative and qualitative analysis of organic or inorganic samples [42–44, 71]. This technique is a useful tool to determine chemical bonding and functional groups of any NMs. FT-IR spectroscopy is based on the absorption of electromagnetic radiation (IR light) with wavelengths within mid-IR region in relation to molecular vibration. The electromagnetic radiation, generated with an IR light source, passes the sample through the interferometer as shown in Fig. 3.14. If a molecule absorbs the IR radiation, the dipole moment is somewhat modified and the molecule becomes IR active. The recorded spectrum shows the absorption or transmission of IR light in dependence on the wavenumber, which is inversely proportional to the wavelength. The obtained positions of absorbing or transmitting peaks reveal molecular vibrations, nature of bonds, and the presence of specific functional groups [42, 90]. Namely, the spectrum exposes the characteristic frequencies of oscillations, from which one can identify the molecule or its part [42, 86, 90].

As was already mentioned, FT-IR spectroscopy is based on IR radiation, which can be absorbed by the sample upon interaction with molecular vibrations. The exact frequency that is required for this interaction depends on numerous factors, such as the type of vibration (stretching, bending, deformation), the strength of the bonds (single, double, triple), and the symmetry of the molecule. Table 3.2 shows some

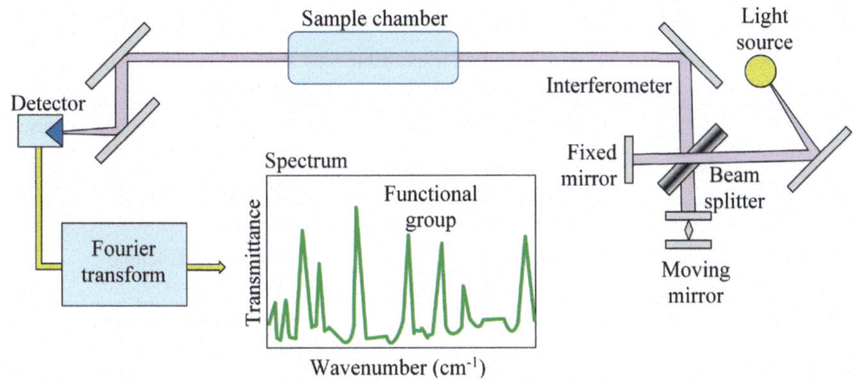

Fig. 3.14 FT-IR spectroscopy

types of vibration and functional groups or compounds corresponding to the region of wavenumbers in the FT-IR spectra [42, 86, 90].

3.2.6 VSM Technique

The vibrating sample magnetometry (VSM) technique is used for the characterization of magnetic NMs as shown in Fig. 3.15. By this technique, a sample is first magnetized in a uniform magnetic field. Then, the sample is sinusoidally vibrated, typically by using a piezoelectric material. The induced voltage in the pickup coil is proportional to the sample's magnetic moment, but does not depend on the strength of the applied magnetic field. In a typical setup, the induced voltage is measured with a lock-in amplifier by using the piezoelectric signal as a frequency reference. It is also possible to record the hysteresis curve of a material by sweeping the magnetic field. The measurements of the magnetic properties by using VSM are necessary to determine the magnetic capacity of metallic NMs [13]. Furthermore, magnetic properties of NMs, such as saturation magnetization M_s, coercivity H_c, and remanence M_r, can also be determined by using VSM [38].

3.3 Synthesis and Characterization of Nanomaterials Used as Fuel Additives

NMs used as fuel additives can be synthesized and characterized by adequate methods by using specific conditions and materials. It is therefore worth to review these processes and conditions briefly, at least for the most frequently used NMs:

Table 3.2 Type of vibration and functional groups for various wavenumbers in FT-IR spectra

Region of wave number (cm^{-1})	Type of vibration responsible for bands in this region	Type of functional group or compound
4000–3200	O–H in N–H stretching	Primary and secondary amines, organic acids, phenols
3335	N–H stretching	Amines
3310–3300	C–H stretching	Alkynes
3100–3000	C–H stretching	Aromatic and olefinic compounds
3000–2800	C–H stretching	Methyl, methylene, and methyne groups
2780	N–CH_2 stretching	Amines
2500–2000	X≡Y, X=Y=Z stretching (X, Y, Z … C, N, O, S)	Alkynes, cyanate, isocyanate, nitrile, isocyanides, azides, diazonium salts, ketenes, thiocyanates, isothiocyanates
1870–1550	C=O stretching	Esters, ketones, amides, carboxylic acids and their salts, acid anhydrides
1690–1620	C=C stretching	Olefinic compounds
1680–1610	N=O stretching	Organic nitrite compounds
1655–1610	– O–NO_2 asymmetric stretching	Organic nitrate compounds
1615	NH_2 in N–H scissoring, bending	Amines
1600–1510	– NO_2 asymmetric stretching	Organic nitro compounds
1600–1450	C=C stretching	Aromatic ring system
1490–1150	H–C–H wagging	Methyl, methylene
1440–1395	C–O stretching	Carboxylic acids
1420–990	S=O stretching	Sulfoxides, sulfates, sulfites, sulfinic acids or esters, sulfones, sulfonic acids, sulfonates, sulfonamides, sulfonyl halides
1360–1250	Aromatic C–N stretching	Amines
1310–1020	C–O–C stretching	Ethers (aromatic, olefinic or aliphatic)
1225–1045	C=S stretching	Thioesters, thioureas, thioamides pyrothiones
1220–1020	Aliphatic C–N stretching	Amines
1000–780	C=C–H deformation	Aliphatic unsaturation
970–840	O–H stretching	Carbocyclic acids
900–670	C–H deformation	Substituted aromatics
850–750	– NH_2 wagging, twisting	Amines
850–500	C–X (X=Cl, Br, I)	Organohalogens

(continued)

Table 3.2 (continued)

Region of wave number (cm^{-1})	Type of vibration responsible for bands in this region	Type of functional group or compound
800–470	Al–O, Al–O–Al, O = Al–OH stretching	Aluminum oxides
785–720	– $(CH_2)_n$, $n = 1, 2, \ldots$	Alkyl groups
750–490	Al–O, Al–O–Al, O=Al–OH stretching	Aluminum oxides
730–480	Zn–O bonds, stretching	Zinc oxides
715	N–H wagging	Amines
700–440	Mg–O stretching	Magnesium oxides
670–560	Co–O stretching	Cobalt oxides
670–525	Ag–O stretching, bending	Silver oxides
630–450	Fe–O stretching	Iron oxides
625–500	Ce–O, Ce–O–Ce stretching	Cerium oxides
620–540	Mn–O, Mn–O–Mn stretching	Manganese oxides
580–520	Cu–O stretching	Copper oxides
525–450	Ni–O stretching	Nickel oxides

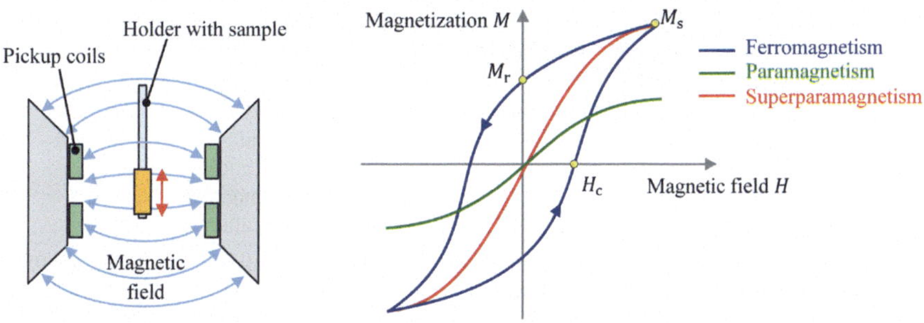

Fig. 3.15 VSM technique

- metallic NPs such as silver (Ag), silver oxide (Ag_2O), aluminum oxide (Al_2O_3), aluminum oxide hydroxide (AlO(OH)), boron oxide (B_2O_3), cerium dioxide (CeO_2), cobalt oxide (Co_3O_4), copper oxide (CuO), maghemite (γ-Fe_2O_3), magnetite (Fe_3O_4), magnesium oxide (MgO), manganese oxide (MnO), dimanganese trioxide (Mn_2O_3), nickel oxide (NiO), silicon dioxide (SiO_2), titanium dioxide (TiO_2), zinc oxide (ZnO), aluminum core and carbon shell (Al@C NPs), and metallic functionalized (CeO_2/mwC(DMF) NTs),
- non-metallic NMs such as carbon nanotubes (C NTs), single-wall carbon nanotubes (swC NTs), multi-wall carbon nanotubes (mwC NTs), graphite oxide nanoparticles (G'O NPs), carbon quantum dot nanoparticles (Cqd NPs), graphene

nanoplatelets (G NPTs), and graphene oxide nanoparticles (GO NPs), and organic NPs, like coconut shell (CS) NPs and rice husk (RH) NPs.

The most important data, such as materials used, synthesis method, conditions, and characterization results for metallic and non-metallic NMs, is assembled in Tables 3.3, 3.4, 3.5, 3.6, 3.7, 3.8, 3.9, 3.10, and 3.11. The most frequently used synthesis methods are: the sol-gel method (Table 3.3), mechanical ball milling method (Table 3.4), coprecipitation method (Table 3.5), sol-gel combustion method (Table 3.6), hydrothermal method (Table 3.7), green synthesis method by using plant extract (Table 3.8), electric arc discharge method (Table 3.9), solvothermal method (Table 3.10), and hybridization by engaging sonochemical method, oxidation, and thermal reduction (Table 3.11). The basic principles of these methods, except of the methods in Table 3.11, which are very simple, have already been summarized in Sect. 3.1. For the sake of convenience, Tables 3.3, 3.4, 3.5, 3.6, 3.7, 3.8, 3.9, 3.10, and 3.11, also list the source literature for each nanomaterial for the case that more detailed data is needed.

Table 3.3 NPs synthesized by the sol-gel method

NMs	Materials	Conditions	Characterization of NMs	References
Ag_2O NPs	Silver nitrate $AgNO_3$, citric acid $C_6H_8O_7$, ethylene glycol $(CH_2OH)_2$, dH_2O	Stirring: 15 min and 3 h Washing: dH_2O Drying: 100 °C, 2 h Calcination: 400 °C, 3 h	XRD: face cubic structure, crystal planes: (111), (200), (220), (311), (222) FT-IR: Ag–O stretching and bending vibration, O–H stretching and bending vibration	[48]
Al_2O_3 NPs	Aluminum nitrate $Al(NO_3)_3$, urea $CO(NH_2)_2$, NaOH, dH_2O	Stirring: 90 °C, 600 rpm, pH 2 Washing: dH_2O Drying: 150 °C, 10 h Calcination: 300 °C, 2 h	TEM: crystalline nature, minimal agglomeration and aggregate formation, average diameter: 20 nm EDXS: Al and O presence XRD: crystalline structure, crystal planes: (110), (200), (211), (310), (411), (521), (541)	[19, 72, 75, 95]
CuO NPs	Copper chloride $CuCl_2$, substrate, dH_2O	Stirring: slowly, 2 h Washing: dH_2O Drying: 100 °C, 24 h Calcination: 600 °C, 6 h	SEM: spherical shape, diameter: 50–210 nm	[49, 68]

(continued)

Table 3.3 (continued)

NMs	Materials	Conditions	Characterization of NMs	References
Mn_2O_3 NPs	Manganese chloride tetrahydrate $MnCl_2 \cdot 4H_2O$, ammonia solution NH_4OH	Stirring: vigorous Washing: MeOH Drying: 120 °C, 12 h Calcination: 400 °C, 2 h	TEM: spherical-like morphology, size: 25–35 nm XRD: cubic bixbyite phase, crystal planes: (211), (222), (400), (332), (431), (440), (541), (622) TGA: mass loss: 18% up to 170 °C due to condensation Mn gel particles, 40% at 400 °C due to methoxide bonded decomposition FT-IR: O–H vibration, Mn–O stretching and bending vibrations	[31]
NiO NPs	Nickel chloride $NiCl_2 \cdot 6H_2O$, Arabic gum solution, dH_2O	Stirring: 75 °C, 120 min Drying: 75 °C, 12 h Calcination: 500 °C, 4 h	SEM: spherical shape XRD: crystalline NPs, size: 14–34 nm, crystal planes (111), (200), (220), and (222), pure cubic phase FT-IR: Ni–O stretching vibration, O–H bonds presence	[29, 89]
SiO_2 NPs	Tetraethyl orthosilicate TEOS, ammonia NH_3, PEG 1000, EtOH, dH_2O	Stirring: 45 °C, 5 h, ultrasonic vibration Drying/Aging: 60 °C, 3 days Calcination: 800 °C, 10 h	SEM: spherical shape, size: 20–30 nm EDXS: presence of Si and O FT-IR: O–H stretching vibration, O–H deformation, Si–O–Si asymmetric and symmetric stretching vibration, Si–O bending vibration	[19, 32, 42]

(continued)

Table 3.3 (continued)

NMs	Materials	Conditions	Characterization of NMs	References
TiO_2 NPs	Titanium tetra isopropoxide $Ti(OCH(CH_3)_2)_4$, HNO_3, EtOH, dH_2O	Stirring: constant rpm, 1 h Drying: 250 °C, 2 h	SEM: size: 30–40 nm XRD: anatase phase, crystal planes: (110), (200), (211) EDXS: Ti and O presence	[70]

Table 3.4 NPs synthesized by the mechanical ball milling method

NMs	Materials	Conditions	Characterization of NMs	References
B_2O_3 NPs	B_2O_3 powder	Milling: 500 rpm, 5 h	SEM: flaky to round shape, average size: 25–95 nm EDXS: B, Au, O presence	[5]
G'O NPs	Graphite flakes	Milling: 600 rpm, 24 h Milling material: stainless steel Ball size: 5 mm	SEM: sphere-like shape EDXS: C (84%), O (14%), N (1.8%), S (0.2%)	[53]
G NPTs	Graphite flakes, melanine	Milling: 100–350 rpm, 30–60 min Mill: planetary, air/nitrogen atmosphere	TEM: nanoplatelets TGA: mass loss due to melanine loss at 300 °C	[50]
CS NPs	Raw coconut shell	Milling: 300 rpm, 5 h	TEM: spherical shape, size: 20 nm EDXS: presence of C, O, Al, Si, K and Ca	[96]
RH NPs	Rice husk	Washing: dH_2O Drying: 24 h Milling: 300 rpm, 5 h	TEM: spherical shape, size: 20 nm EDXS: presence of C, O, Mg, Si and K	[97]

Table 3.3 shows that the sol-gel method can be used for various metallic NPs, such as Ag_2O, Al_2O_3, CuO, Mn_2O_3, NiO, SiO_2, and TiO_2.

The Ag_2O NPs synthesis is described by Kumar and Manisha [48]. At first, solutions of silver nitrate $AgNO_3$ and citric acid $C_6H_8O_7$ are stirred continuously for 15 min. After addition of ethylene glycol $(CH_2OH)_2$ and mixing for 3 h, the obtained NPs are washed with dH_2O, dried in an oven at 100 °C for 2 h, and calcined in a muffle furnace at 400 °C for 3 h.

Table 3.5 NPs synthesized by the coprecipitation method

NMs	Materials	Conditions	Characterization of NMs	References
CuO NPs	Copper chloride $CuCl_2$, copper nitrate $Cu(NO_3)_2 \cdot 3H_2O$, NaOH, dH_2O, EtOH	Stirring: slowly Precipitation: pH 14 Washing: dH_2O, EtOH Drying: 80 °C, 16 h Calcination: 500 °C, 4 h	SEM: spherical shape, diameter: 50 nm XRD: crystal planes: (110), (002), (200), (202) FT-IR: Cu–O stretching vibration, O–H bending vibrations	[69]
γ-Fe_2O_3 NPs	Ferrous chloride tetrahydrate $FeCl_2{\cdot}4H_2O$, ferric chloride hexahydrate $FeCl_3{\cdot}6H_2O$, ammonium solution NH_4OH, dH_2O	Stirring: 87 °C, 300 rpm Precipitation: 60 min, pH 9.9 Washing: dH_2O, several times Drying: 80 °C, 10 h	TEM: spherical shapes EDXS: presence of Fe and O XRD: crystal planes (220), (311), (400), (422), (511), (440), (620), and (533), lattice parameter 0.8358 nm, interplanar spacing 0.252 nm TGA: mass loss: 1.7% due to absorbed moisture up to 160 °C, 1.3% due to reaction γ-$Fe_2O_3 \rightarrow Fe_2O_3 \rightarrow 2FeO$ FT-IR: presence of Fe–O and O–H bonds VSM: mass saturation magnetization of 53.79 emu/g, remanence of 5.25 emu/g, coercitivity of 202.90 Oe	[42–44, 57]
Fe_3O_4 NPs	Ferrous chloride tetrahydrate $FeCl_2 \cdot 4H_2O$, ferric chloride hexahydrate $FeCl_3 \cdot 6H_2O$, ammonium solution NH_4OH, dH_2O	Stirring: 70 °C, 370 rpm, 30 min Precipitation: 70 °C, pH 9.9, 30 min Washing: dH_2O, three times Drying: 100 °C	TEM: spherical shapes, monodisperse NPs EDXS: presence of Fe and O XRD: crystal planes (220), (311), (400), (422), (511), (440), and (731) FT-IR: presence of Fe–O and O–H bonds VSM: mass saturation magnetization of 46.7 emu/g, remanence of 1.65 emu/g, coercitivity of 0 Oe	[9, 30, 77, 79]

Table 3.6 NPs synthesized by the sol-gel combustion method

NMs	Materials	Conditions	Characterization of NMs	References
CeO_2 NPs	Cerium nitrate hexahydrate $Ce(NO_3)_3 \cdot 6H_2O$, glycine $C_2H_5NO_2$, dH_2O	Stirring: 60–70 °C, 120–150 min Combustion: 100 °C Annealing: 600–650 °C, 2–3 h	SEM: large agglomeration with high porous nature EDXS: Ce and O presence XRD: pure cubic fluorite structure, crystal planes: (111), (200), (220), (311), (222), (400), (331), (420)	[7, 24]
MgO NPs	Magnesium glycinate $C_4H_8MgN_2O_4$, magnesium nitrate $Mg(NO_3)_2$, dH_2O	Stirring: room temperature, 60 min Heating: 160 °C, 60 min Combustion: 170 °C Calcination: 600 °C, 2 h	XRD: crystalline NPs, size: 21 nm FT-IR: Mg–O stretching vibration, O–H stretching and bending vibration	[61, 80]
ZnO NPs	Zinc nitrate hexahydrate $Zn(NO_3)_2 \cdot 6H_2O$, cassava starch, dH_2O	Stirring and heating: 200 °C Combustion: 475 °C, 5 min Calcination: 500 °C, 5 h	XRD: hexagonal wurtzite symmetry, crystal planes: (100), (002), (101), (102), (110), (103), (200), (112), (201) FT-IR: O–H stretching vibration of Zn–OH–ZN, CO, OH, Zn–O–Zn, Zn–O–O–Zn presence, Zn–O stretching vibration	[78]

Al_2O_3 NPs synthesis is described by Radhakrishnan et al. [72] and by Venu and Madhavan [95]. At first, 0.5 M aluminum nitrate solution $Al(NO_3)_3 \cdot 9\,H_2O$ is stirred magnetically at 22 °C. Then, 0.05 M urea solution is added to the aluminum nitrate solution, and the obtained solution is stirred at 90 °C at 600 rpm until pH 2 is reached. The prepared solution is titrated with 0.1 M NaOH solution until pH 6 is reached. Further addition of NaOH until one gets pH 8, forms a gel. After drying at 150 °C for up to 10 h and calcination at 300 °C for 2 h, the Al_2O_3 NPs are collected.

CuO NPs synthesis is described by Lenin et al. [49]. A mechanical mixing of hydrated salt is done by using an ultrasonic cleaner device for 2 h. After drying in a hot-air oven at 100 °C for 24 h, the obtained fine powder is kept in muffle furnace at 600 °C for 6 h to get the required CuO NPs.

Mn_2O_3 NPs synthesis is described in Gnanam and Rajendran [31]. After stirring a solution of manganese chloride tetrahydrate $MnCl_2 \cdot 4H_2O$, aqueous ammonia

Table 3.7 NPs synthesized by the hydrothermal method

NMs	Materials	Conditions	Characterization of NMs	References
CeO_2 NPs	Cerium chloride $CeCl_3$, ammonium hydroxide NH_4OH, sodium hydroxide NaOH, dH_2O, Milli-Q water	Stirring: 300 rpm, 30 min Heating: 300 °C, 1 h Autoclave: 150 °C, 48 h Washing: Milli-Q water Drying: 60 °C, in hot air Annealing: 800 °C, 2 h	SEM: needle-like morphology with lots of pores EDXS: Ce and O presence XRD: face-centered cubic structure, crystal planes (111), (200), (220), (311), (222), (400), (331), (420)	[94]
Co_3O_4 NPs	Cobalt chloride $CoCl_2$, triton X-100, KOH, dH_2O, EtOH	Autoclave: 180 °C, 6 h Cooling: to room temperature Washing: dH_2O, EtOH Drying: in vacuum, 90 °C, 6 h	TEM: spherical shape XRD: cubic unit cell, spinel structure, crystal planes: (111), (220), (311), (222), (400), (422), (511), (440) FT-IR: Co–O stretching vibration, O–H stretching and binding vibration VSM: paramagnetic behavior, high purity	[104]
Cqd NPs	Orange peel, dH_2O	Autoclave: 180 °C, 12 h Cooling: to room temperature Centrifugation: 5000 rpm Drying: 80 °C	TEM: dot morphology, mean size: 1–3 nm, spheroidal shaped NPs XRD: crystal plane (002) FT-IR: O–H, C–H, C=C, and C–O bonds	[28]

solution NH_4OH is added dropwise under stirring. The resulting opal gels are filtered and washed with MeOH to remove impurities, and dried at 120 °C for 12 h in order to remove water molecules and get white colored powder. Finally, black-brown colored Mn_2O_3 NPs are formed at 400 °C after about 2 h.

NiO NPs synthesis is described by Fardood et al. [29]. A nickel chloride $NiCl_2 \cdot 6H_2O$ solution is added to *Arabic gum* solution, previously prepared by stirring for 120 min at 75 °C. A container with the obtained solution is placed into a sand bath of 75 °C and stirred for 12 h. The final product is calcined at 500 °C in air for 4 h to obtain NiO NPs.

SiO_2 NPs obtained by the sol-gel method are synthesized via hydrolysis and condensation of tetraethyl orthosilicate (TEOS), which is catalyzed by ammonia according to [32, 42]:

Table 3.8 NPs synthesized by the green method using plant extract

NMs	Materials	Conditions	Characterization of NMs	References
Ag NPs	Silver nitrate $AgNO_3$, extract of *Shikakai* and *Reetha*, NH_4OH, Milli-Q water	Centrifugation: 10,000 rpm, 25 min Centrifugation: 15,000 rpm, 30 min Washing: dH_2O	TEM: monodisperse, spherical shape; average diameter: 20–40 nm XRD: face-centered cubic silver structure, crystal planes: (111), (200), (220), (311)	[62, 92]
Ag_2O NPs	Silver nitrate $AgNO_3$, ethanolic extract of *Pavetta indica Linn*, dH_2O	Stirring: 100–120 °C, 24 h Ultracentrifugation: 5000–10,000 rpm Drying: 140 °C, 8 h Calcination: 450 °C, 3 h	SEM: distorted square shape, 50 nm size, 170 nm width EDXS: Ag and O presence	[93]
MnO NPs	Manganese acetate $Mn(CH_3CO_2)_2 \cdot xH_2O$, extract of *Dittrichia graveolens*, extract of *Curcumin*, EtOH, dH_2O	Centrifugation: 3500 rpm, 15 min Washing: EtOH, dH_2O	SEM: average diameter: 38 nm FT-IR: O–H stretching vibration, C=C, C=O, C–O bending modes	[88]

$$Si(OC_2H_5)_4 + H_2O \rightarrow Si(OC_2H_5)_3OH + C_2H_5OH$$

$$\equiv Si{-}O{-}H + H{-}O{-}Si \equiv \rightarrow \equiv Si{-}O{-}Si \equiv + H_2O$$

$$\equiv Si{-}O{-}C_2H_5 + H{-}O{-}Si \equiv \rightarrow \equiv Si{-}O{-}Si \equiv + C_2H_5OH$$

In a typical procedure, described by Guo et al. [32], TEOS is first added dropwise into a homogenous mixture of EtOH and dH_2O. At the same time, a certain amount of PEG 1000 acting as a surfactant is added. After sonication of the mixture for 2 h at 45 °C, a mixture of ammonia and EtOH is added and the mixture is then agitated for further 3 h under ultrasonic vibration. This is followed by aging for three days at 60 °C in a vacuum box (without ultrasonication) to obtain a SiO_2 gel. After drying in a vacuum box at 80 °C, the obtained SiO_2 NPs are calcined in air at 800 °C for 10 h.

In the preparation of TiO_2 NPs, described by Praveen et al. [70], the titanium tetra isopropoxide $Ti(OCH(CH_3)_2)_4$ is added into a mixture of EtOH and dH_2O under constant stirring for 1 h. Next, nitric acid HNO_3 is added into this solution to obtain the sol-gel, which is heated at 250 °C for 2 h in a furnace to evaporate the water and increase the viscosity of the solution.

Table 3.4 shows that the mechanical ball milling method can be used for B_2O_3 NPs, G'O NPs, G NPTs, CS NPs, and RH NPs.

Table 3.9 NPs synthesized by the electric arc discharge method

NMs	Materials	Conditions	Characterization of NMs	References
Al@C NPs	Graphite powder, aluminum powder, EtOH	Stirring: 1 h Drying: 80 °C, 2 days Current voltage: 25 V Current power supply: 125 A Arc application: a few min	SEM: spherical shape, size: 60–160 nm TEM: core@shell structure EDXS: presence of Al, C and Si XRD: crystal planes of alumina (111), (200), (220), (311), and (222) corresponding to face-centered cubic crystal structure and crystal plane of hexagonal graphite structure (002)	[18, 100]
C NTs; swC NTs; mwC NTs	Graphite, dH_2O, catalyst: Fe, Co …	Current voltage: 20–30 V Current power supply: 100–200 A Arc application: a few min	SEM: nanotubes, average diameter: 15–115 nm EDXS: C and O presence XRD: crystal plane: (002)	[6, 12, 19, 85]

B_2O_3 NPs synthesis is described by Alizadeh et al. [5]. To reduce the particles size of B_2O_3 powder from micron level to nanosize, a high-energy planetary ball mill in a chromium steel chamber under a high-purity argon atmosphere is used with chromium carbide balls of 10 mm diameter. B_2O_3 NPs are obtained by milling after about 5 h with rotation of the planet at 500 rpm and with ball-to-powder weight ratio of 10:1.

G'O NPs preparation is described by Mahmoud et al. [53]. The dry ball milling process is carried out in a planetary ball mill under ambient conditions. The ball size is 5 mm for both grinding materials, stainless steel and zirconia. The rotation frequency is maintained at 600 rpm for the milling time of up to 24 h.

G NPTs synthesis is described in León et al. [50]. Exfoliate graphite and 2,4,6-triamine-1,3,5-triazine (melanine) are ball-milled at 100–350 rpm in air or nitrogen atmosphere for 30–60 min in a planetary mill.

CS NPs preparation is described in Vinukumar et al. [96]. Dry raw coconut shell is hammered, sieved, mixed, and then put into the planetary mono mill. The mechanical milling process is performed for 5 h at a rotating speed of 300 rpm. Then, grounded CS NPs are dried at room temperature.

Table 3.10 NPs synthesized by the solvothermal method

NMs	Materials	Conditions	Characterization of NMs	References
Ag NPs	Silver nitrate $AgNO_3$, sodium borohydride $NaBH_4$, EtOH, dH_2O	Stirring/Heating: by reflux, 80 °C, 40 h Centrifugation: 8000 rpm, 20 min Washing: dH_2O Drying: in oven, 60 °C, 2 h	TEM: monodisperse, spherical shape, smooth surface, highly uniform in size, grain size: 5 nm XRD: face-centered cubic silver structure, crystal planes: (111), (200), (220)	[99]
AlOOH NPs	Aluminum nitrate $Al(NO_3)_3 \cdot 9H_2O$, aluminum chloride $AlCl_3$, EtOH, NaOH, dH_2O	Stirring: 10–15 min, room temperature Autoclave: 180 °C, 24 h Centrifugation Washing: EtOH, dH_2O Drying: 65 °C, over night	SEM: spherical-shaped NPs, average diameter: <50 nm XRD: crystalline structure, crystal planes: (020), (021), (041), (002), (200) FT-IR: O=Al–(OH) angle deformation, (OH)–Al=O angle bending, Al–O–Al stretching vibration, angle bending and deformation of H-bonds of AlO(OH), NO_3^- presence, O–H stretching and bending	[2, 81]

RH NPs synthesis is described in Vinukumar et al. [97]. The collected rice husk is washed with dH_2O and dried for 24 h. By using a mixer, the dried material is then grounded. The rice husk is milled for about 5 h with a rotational speed of 300 rpm. The ground powder is finally dried at room temperature.

Table 3.5 shows that the coprecipitation method can be used for metallic NPs, such as CuO, γ-Fe_2O_3, and Fe_3O_4.

The synthesis of CuO NPs is described in Phiwdang et al. [69]. The NaOH solution is added dropwise to a mixture of copper chloride $CuCl_2$ and copper nitrate $Cu(NO_3)_2 \cdot 3H_2O$ solutions under vigorous stirring until pH 14 is reached. The obtained black precipitates are then washed by dH_2O and EtOH for several times and dried at 80 °C for 16 h. Finally, the obtained CuO NPs are calcined at 500 °C for 4 h.

γ-Fe_2O_3 NPs synthesis is described in Kegl [42] and in Kegl et al. [45]. γ-Fe_2O_3 NPs can be synthesized from ferrous chloride tetrahydrate $FeCl_2 \cdot 4H_2O$ and ferric chloride hexahydrate $FeCl_3 \cdot 6H_2O$, which are added to ammonium solution NH_4OH

Table 3.11 NMs synthesized by the hybridization method, oxidation, and thermal reduction

NMs	Materials	Conditions	Characterization of NMs	References
CeO_2/mwC (DMF) NTs	CeO_2 NPs, mwC NTs, hydrogen peroxide H_2O_2, thionyl chloride $SOCl_2$, dH_2O, EtOH, dimethylformamide (DMF), anhydrous tetrahydrofuran (THF), MeOH, octadecylamine	Calcination: 460 °C, 3 h Ozonation: AC, 4 h Ozonation: 60 °C, 1 h, 36% H_2O_2 Stirring: RT, 20 h Washing/Drying: MeOH/100 °C Stirring: 70 °C, 24 h Washing/Drying: THF/40 °C, 30 min Sonication: 60 °C, 40 GHz, 2 h Cooling/Washing: RT, EtOH Drying: 70 °C, overnight	SEM: tubes, average diameter: 40–50 nm FT-IR: OH stretching vibration, C=C bond of mwC NTs, Ce–O–C stretching vibration, Ce–O stretching bond	[54]
G'O NPs	Graphite flakes, $NaNO_3$, H_2SO_4, $KClO_3$, dH_2O	Mixing: 20 h Washing and centrifugation: dH_2O Drying: 50 °C, 72 h, in vacuum	TGA: mass loss of 7% due to absorbed moisture at 25–100 °C, and 20% due to evolution of CO, CO_2 and H_2O at 100–350 °C FT-IR: presence of O–H, C=O, C=C, and C–O functional groups	[39, 40, 65]
GO NPs	Graphite oxide, N_2	Heating: 90 °C Nitrogen flow: 50 mL/min Residence time: 5 min	SEM: layered structure EDXS: presence of C, O, S XRD: crystalline structure FT-IR: C–O, C=C, and O–H bonds	[25, 27, 33, 67, 105]

and heated by reflux to 87 °C at 300 rpm. The reaction is carried out in a round bottom reaction flask for 60 min at pH 9.9. The particles are precipitated from the reaction mixture by a permanent magnet, washed several times with dH_2O, and finally dried at 80 °C for 10 h.

Fe_3O_4 NPs preparation is described in Ramanan and Yuvarajan [77] and in Ashour et al. [9]. At first, $FeCl_2 \cdot 4H_2O$ and $FeCl_3 \cdot 6H_2O$ are dissolved in dH_2O by slowly increasing the temperature to 70 °C under nitrogen atmosphere with mechanical stirring at 370 rpm for 30 min. Then, ammonium hydroxide solution NH_4OH is added instantaneously to the reaction mixture and left for another 30 min at 70 °C. The obtained black precipitates are washed with dH_2O three times and separated from the supernatant by using a permanent magnet to obtain Fe_3O_4 NPs.

Table 3.6 shows that the sol-gel combustion method can be used for metallic NPs, such as CeO_2 NPs, MgO NPs, and ZnO NPs.

CeO_2 NPs synthesis is described by Annamalai et al. [7] and by Dhinesh and Annamalai [24]. By the most frequently used sol-gel combustion method, cerium nitrate hexahydrate $Ce(NO_3)_3 \cdot 6H_2O$ (oxidizing reactant) and glycine $C_2H_5NO_2$ (reducing agent) are dissolved in dH_2O. The solution is stirred for 2 h at 60 °C and then transferred to an electric mantle at 100 °C to form a transparent viscous gel by dehydration of excess of H_2O. After some time, the gel undergoes a strong self-propagating combustion reaction to yield a yellow porous foam of NPs. A large amount of non-toxic gases like N_2, CO_2, and H_2O are released in the form of brown fumes during the whole combustion reaction. The obtained CeO_2 NPs are annealed at 600 °C for 3 h.

MgO NPs preparation by the sol-gel combustion method is described in Ranjan et al. [80] and in Nassar et al. [61]. Magnesium glycinate $C_4H_8MgN_2O_4$ and magnesium nitrate $Mg(NO_3)_2$ are dissolved in dH_2O. The solution is stirred at room temperature for 60 min and heated at 160 °C for 60 min. Magnetic pellet is removed from the solution when the liquid becomes gel. Foam-like substance is formed on further heating until combustion of this substance takes place at a temperature of 170 °C. The particles formed during combustion are collected and ground into fine particles using a mortar and pestle. Calcination is performed for 2 h at 600 °C to obtain MgO NPs.

ZnO NPs synthesis is described by Ramasami et al. [78]. Zinc nitrate hexahydrate $Zn(NO_3)_2 \cdot 6H_2O$ and *Cassava starch*, acting as a combustion agent, are mixed well with dH_2O and heated on a hot plate at 200 °C until a viscous gel is formed. The gel is introduced into a muffle furnace at around 475 °C, where the combustion reaction occurs within 5 min with the evolution of gases like NO_2, CO_2, and water vapor. The self-propagating heat wave, resulting in a smoldering type of reaction, occurs with a flaky voluminous product formation. The ZnO NPs are calcined at 500 °C for 5 h to remove the impurities.

Table 3.7 shows that the hydrothermal method can be used for CeO_2 NPs, Co_3O_4 NPs, and Cqd NPs.

CeO_2 NPs synthesis is described by Vairamuthu et al. [94]. Cerium chloride $CeCl_3$ solution is prepared and stirred at 300 rpm. This solution is then enriched with the ammonium hydroxide NH_4OH solution and further stirred at 300 rpm for

30 min. After heating this solution at first at 110 °C and then at 300 °C for 1 h in a muffle furnace, the powder is transferred to an autoclave containing sodium hydroxide NaOH and is then heated at 150 °C for 48 h. After that the system is allowed to cool down to room temperature. The resulting solution is washed with Milli-Q water to get rid of the salts. After drying in hot air at 60 °C, the obtained powder is annealed at 800 °C for 2 h.

Co_3O_4 NPs preparation is described by Jeryrajkumar et al. [37] and by Yarestani et al. [104]. The reaction mixture of cobalt chloride $CoCl_2$ solution, triton X-100 $C_{14}H_{22}O(C_2H_4O)_n$ as surfactant, and potassium hydroxide KOH is put into an autoclave at 180 °C for 6 h. After cooling the autoclave to room temperature, a dark precipitate is obtained, which is filtered, washed with dH_2O and EtOH, and dried at 90 °C for 6 h under vacuum.

Cqd NPs preparation by hydrothermal method by using orange peel is described by Ettefaghi et al. [28]. At first, orange peel is powdered, mixed with dH_2O, and put into an autoclave. This solution is kept under hydrothermal conditions for 12 h at 180 °C. After cooling the autoclave body to room temperature, the precipitates are separated from the solution by centrifugation at 5000 rpm. These precipitates are dried at 80 °C so that Cqd NPs are obtained.

Table 3.8 shows that the green method using plant extract can be used for Ag, Ag_2O, and MnO NPs.

Ag NPs preparation is described by Sur et al. [92]. After the preparation of aqueous plant extract of *Reetha* and *Shikakai*, this extract is added to the mixture of $AgNO_3$ and ammonia NH_4OH solution. After centrifugation at 10,000 rpm for 25 min, followed by centrifugation of collected supernatant at 15,000 rpm for 30 min, and washing the pellets with Milli-Q water, the Ag NPs are obtained.

Ag_2O NPs synthesis is described in Suresh et al. [93]. After magnetic stirring of silver nitrate $AgNO_3$ solution with the plant extract of *Pavetta indica Linn* within the temperature range of 100–120 °C for about 24 h, metal ions form a complex with phytoconstituents of the plant extract. The formed complex is ultracentrifuged at 5000–10,000 rpm for 20 min. After drying of the residual complex in an air oven at 140 °C, the complex is calcined in a muffle furnace at 450 °C for about 3 h.

MnO NPs preparation is described by Souri et al. [88]. A solution of manganese acetate $Mn(CH_3CO_2)_2 \cdot xH_2O$ and extract of *Dittrichia graveolens* is prepared and mixed. After the solution rests for up to 120 min, a fresh extract of *Curcumin* is added to the solution. MnO NPs are separated from the solution by centrifugation at 3500 rpm for 15 min and washed several times by using EtOH and dH_2O.

Table 3.9 shows that the electric arc discharge method can be used for the synthesis of core@shell Al@C NPs, C NTs, swNTs, and mwC NTs.

Al@C NPs synthesis is described by Chen et al. [18] and by Zhang et al. [106]. A mixture of graphite powder, aluminum powder, and EtOH is stirred for 1 h under ultrasonic vibration, dried at 80 °C for 2 days after filtering out the EtOH, followed by pressing to form a cylindrical disk that serves as the anode. A cylindrical graphite rod is applied as the cathode. The chamber is evacuated, and Ar with pressure of 1 Pa is introduced after the anode and cathode are placed into the chamber. To produce the arc between the cathode and the anode disk, a current of 125 A at 25 V is used. After rapid vaporization at high temperature and the subsequent cooling, carbon-coated metal NPs are obtained.

C NTs, swC NTs, and mwC NTs preparation is described in Sharma et al. [85], Basha and Anand [12], and Ando et al. [6]. When pure graphite rods are used, the anode evaporates to form fullerenes, which are deposited in form of soot in the chamber. However, a small part of the evaporated anode is deposited on the cathode, which includes C NTs. These C NTs, made of coaxial graphene sheets and called multi-wall carbon nanotubes (mwC NTs), are found not only on the top surface of the cathode deposit but also deep inside the deposit. Large-scale synthesis of mwC NTs by arc discharge has been achieved in He gas, and their thermal purification has also been successful. When a graphite rod containing metal catalyst (Fe, Co, etc.) is used as the anode in conjunction with a pure graphite cathode, single-walled carbon nanotubes (swC NTs) are generated in form of soot [6]. For example, mwC NTs are synthesized by two graphite rods of 7 and 20 mm in diameter, which are used as anode and cathode electrodes, respectively [85]. The arc between the electrodes is established by a direct current power supply capable to provide 100–200 A current at a voltage range of 20–30 V. The reaction environment is provided by an open vessel containing dH_2O. By varying the current and voltage values, the produced mwC NTs may exhibit diameters in the range of 15–150 nm [85]. After the arc application time of a few minutes, the discharge is stopped and cathodic deposit, which contains C NTs along with the soot, is collected from the chamber walls. The deposit is further purified and observed under an electron microscope to investigate its morphology [12].

Table 3.10 shows that the solvothermal method can be used for Ag NPs and AlOOH NPs.

Ag NPs preparation is described by Wani et al. [99]. A reaction mixture of silver nitrate $AgNO_3$ solution, sodium borohydrate $NaBH_4$ acting as reducing agent, and EtOH as refluxing reagent, are stirred by a magnetic stirrer accompanied by refluxing at 80 °C for 40 h. After centrifugation at 8000 rpm for 20 min, the supernatant is decanted. The precipitate is washed with dH_2O three times to remove the solvents, if any, and dried in an oven maintained at 60 °C for 2 h.

AlO(OH) NPs synthesis is described by Abdollahifar et al. [2]. An aluminum nitrate solution $Al(NO_3)_3 \cdot 9\,H_2O$ or aluminum chloride $AlCl_3$ is dissolved in EtOH and stirred for 10–15 min at room temperature. 2 M NaOH solution is subsequently added dropwise to the solution to give lacteous precipitates. The reaction mixture at pH 5 or pH 11 is transferred into an autoclave and then placed into an oven at 180 °C for 24 h. The obtained samples are then treated by centrifugation, washed with EtOH and dH_2O several times, and dried overnight at 65 °C in an oven.

Hybrid NMs from cerium oxide and multi-wall carbon nanotubes, functionalized by *N*,*N*-dimethylformamide (CeO_2/mwC(DMF) NTs), given in Table 3.11, are synthesized by hybridization of CeO_2 NPs with mwC NTs by using the sonochemical method. Table 3.11 contains also some data related to graphite oxide nanoparticles (G'O NPs), synthesized by oxidation of graphite flakes, and graphene oxide nanoparticles (GO NPs), synthesized by thermal reduction of graphite oxide.

CeO_2/mwC(DMF) NTs are synthesized by hybridization of previously prepared CeO_2 with mwC NTs [54]. Calcination of CeO_2-mwC NTs is done at 460 °C for 3 h. Hybridized CeO_2/mwC NTs are then carboxyl-functionalized sequentially by the first ozonation at room temperature and atmospheric condition for 4 h and then by the second ozonation by using 36% hydrogen peroxide H_2O_2 at 60 °C for 1 h. After that the mixture is stirred at room temperature for 20 h. Subsequently, the solution is passed through a 0.2 mm polycarbonate membrane, washed with MeOH, and oven-dried at 100 °C. The obtained solution is stirred and refluxed in a mixture of thionyl chloride $SOCl_2$ and dimethylformamide (DMF) at 70 °C for 24 h. After acylchlorination, the hybridized nanocatalyst is washed with anhydrous tetrahydrofuran (THF) for five times and then dried at 40 °C for 30 min. The acyl-chlorinated nanocatalyst is then added to octadecylamine (oda). This mixture is sonicated using an ultrasound bath at 60 °C at 40 GHz for 2 h, and then refluxed for 2 days. After cooling to room temperature, the amidated product is washed with EtOH to remove excess amine. Finally, the nanocatalyst containing CeO_2 and amides' functional groups are dried at 70 °C overnight.

G'O NPs synthesis by oxidation of graphite flakes is described by Kartick et al. [40] and Ooi et al. [65]. Graphite flakes are oxidized with a mixture of sodium nitrate $NaNO_3$, concentrated sulfuric acid H_2SO_4, and potassium chlorate $KClO_3$. Then, dH_2O is added and after 20 h of reaction a dark brown product is obtained, which is washed and centrifuged several times with dH_2O and finally dried in a vacuum oven at 50 °C for 72 h.

GO NPs preparation by thermal reduction of graphite oxide is described in Drewniak et al. [25], El-Seesy et al. [27], Zaaba et al. [105], Gupta et al. [33], and by Paulchamy et al. [67]. GO NPs are obtained by thermal treatment of graphite oxide at 90 °C in a vertical furnace under a nitrogen flow of 50 mL/min. The residence time at the final temperature is 5 min.

References

1. Abdelhamid HN, Zou X (2018) Template-free and room temperature synthesis of hierarchical porous zeolitic imidazolate framework nanoparticles and their dye and CO_2 sorption. Green Chem 20(5):1074–1084
2. Abdollahifar M, Hidaryan M, Jafari P (2018) The role anions on the synthesis of AlOOH nanoparticles using simple solvothermal method. Bol Soc Esp Cerám Vidrio 57:66–72
3. Accardo G, Frattini D, Ham HC, Han JH, Yoon SP (2018) Improved microstructure and sintering temperature of bismuth nano-doped GDC powders synthesized by direct sol-gel combustion. Ceram Int 44:3800–3809
4. Ali A, Zafar H, Zia M, Haq I, Phuli AR, Ali JS, Hussain A (2016) Synthesis, characterization, applications, and challenges of iron oxide nanoparticles. Nanotech Sci Appl 9:49–67
5. Alizadeh M, Sharifianjazi F, Haghshebasjazi E, Aghakhani M, Rajabi L (2015) Production of nanosized boron oxide powder by high-energy ball milling. Synth React Inorg Met-Org Nano-Met Chem 45(1):11–14
6. Ando Y, Zhao X, Sugai T, Kumar M (2004) Growing carbon nanotubes. Mater Today 7(10):22–29

7. Annamalai M, Dhinesh B, Nanthagopal K, Sivaramakrishnan P, Lalvani JIJ, Parthsarathy M, Annamalai K (2016) An assessment on performance, combustion and emission behavior of a diesel engine powered by ceria nanoparticle blended emulsified biofuel. Energy Convers Manag 123:372–380
8. Arora N, Sharma NN (2014) Arc discharge synthesis of carbon nanotubes: comprehensive review. Diam Relat Mater 50:135–150
9. Ashour RM, Abdel-Khalek AA, Ali MM, Abdel-Magied AF (2016) Adsorption of La^{3+} and Gd^{3+} using magnetic iron oxide nanoparticles: mechanistic and kinetic study. Chem Technol 11(3):100–108
10. Asim N, Ahmadi S, Alghoul MA, Hammadi FY, Saeedfar K, Sopian K (2014) Research and development aspects on chemical preparation techniques of photoanodes for dye sensitized solar cells. Int J Photoenergy 518156:1–21
11. Awel S, Kirian RA, Wiedorn MO, Beyerlein KR, Roth N, Horke DA, Oberthür D, Knoska J, Mariani V, Morgan A, Adriano L, Tolstikova A, Xavier PL, Yefanov O, Aquila A, Barty A, Roy-Chowdhury S, Hunter MS, James D, Robinson JS, Weierstall U, Rode AV, Bajt S, Küppera J, Chapmana HN (2018) Femtosecond X-ray diffraction from an aerosolized beam of protein nanocrystals. J Appl Crystallogr 51(1):133–139
12. Basha JS, Anand RB (2014) Performance, emission and combustion characteristics of a diesel engine using carbon nanotubes blended Jatropha methyl ester emulsions. Alex Eng J 53:259–273
13. Basualto C, Gaete J, Molina L, Valenzuela F, Yañez C, Marco JF (2015) Lanthanide sorbent based on magnetite nanoparticles functionalized with organophosphorus extractants. Sci Technol Adv Mater 16:1–9
14. Bellardita M, Paola AD, Megna B, Palmisano L (2018) Determination of the crystallinity of TiO_2 photocatalysts. J Photochem Photobiol A 367:312–320
15. Bhagyaraj SM, Oluwafeni OS, Kalarikkal N, Thomas S (2018) Characterization of nanomaterials. Advances and key technologies. Elsevier, Duxford
16. Bhavani P, Rajababu CH, Arif MD, Reddy IVS, Reddy NR (2017) Synthesis of high saturation magnetic iron oxide nanomaterials via low temperature hydrothermal method. J Magn Magn Mater 426:459–466
17. Cargnello M, Gordon TR, Murray CB (2014) Solution-phase synthesis of titanium dioxide nanoparticles and nanocrystals. Chem Rev 114:9319–9345
18. Chen Y, Luo W, Wang J, Huang J (2017) Enhanced thermal conductivity and durability of a paraffin wax nanocomposite based on carbon-coated aluminum nanoparticles. J Phys Chem C 121:12603–12609
19. Chen AF, Adzmi MA, Adam A, Othman MF, Kamaruzzaman MK, Mrwan AG (2018) Combustion characteristics, engine performances and emissions of a diesel engine using nanoparticle-diesel fuel blends with aluminium oxide, carbon nanotubes and silicon oxide. Energy Convers Manag 171:461–477
20. Danks AE, Hall SR, Schnepp Z (2016) The evolution of 'sol-gel' chemistry as a technique for materials synthesis. Mater Horiz 3:91–112
21. Darr JA, Zhang J, Makwana NM, Weng X (2017) Continuous hydrothermal synthesis of inorganic nanoparticles: applications and future directions. Chem Rev 117:11125–11238
22. Dhand C, Dwiwedi N, Loh XJ, Ying ANJ, Verma NK, Beuerman RW, Lakshminarayanan R, Ramakrishna S (2015) Methods and strategies for the synthesis of diverse nanoparticles and their applications: a comprehensive overview. RSC Adv 5:105003–105037
23. Dharuman S, Venkata SBS, Kandasamy SK (2018) Effects on diesel engine characteristics and emission of nanofluids blends with diesel. Int J Chem Tech Res 11(2):418–426
24. Dhinesh B, Annamalai M (2018) A study on performance, combustion and emission behaviour of diesel engine powered by novel nano nerium oleander biofuel. J Clean Prod 196:74–83
25. Drewniak S, Muzyka R, Stolarczyk A, Pustelny T, Kotyczka-Morańska M, Setkiewicz M (2018) Studies of reduced graphene oxide and graphite oxide in the aspect of their possible application in gas sensors. Sensors 16(1):1–16

26. Duan H, Wang D, Li Y (2015) Green chemistry for nanoparticle synthesis. Chem Soc Rev 44(16):5778–5792
27. El-Seesy AI, Hassan H, Ookawara S (2018) Performance, combustion, and emission characteristics of a diesel engine fueled with Jatropha methyl ester and graphene oxide additives. Energy Convers Manag 166:674–686
28. Ettefaghi E, Ghobadian B, Rashidi A, Najafi G, Khoshtaghaza MH, Rashtchi M, Sadeghian S (2018) A novel bio-nano emulsion fuel based on biodegradable nanoparticles to improve diesel engines performance and reduce exhaust emissions. Renew Energy 125:64–72
29. Fardood ST, Ramazani A, Moradi S (2017) A novel green synthesis of nickel oxide nanoparticles using arabic gum. Chem J Moldova Gen Ind Ecol Chem 12(1):115–118
30. Ghandoor HE, Zidan HM, Khalil MMH, Ismail MIM (2012) Synthesis and some physical properties of magnetite (Fe_3O_4) nanoparticles. Int J Electrochem Sci 7:5734–5745
31. Gnanam S, Rajendran V (2011) Synthesis of CeO_2 or α-Mn_2O_3 nanoparticles via sol-gel process and their optical properties. J Sol-Gel Sci Technol 58:62–69
32. Guo Q, Huang D, Kou X, Cao W, Li L, Ge L, Li J (2017) Synthesis of disperse amorphous SiO_2 nanoparticles via sol-gel process. Ceram Int 43:192–196
33. Gupta DK, Rajaura RS, Sharma K (2015) Synthesis and characterization of graphene oxide nanoparticles and their antibacterial activity. SGVU Int J Env Sci Technol 1(1):16–24
34. Hinman JJ, Suslik KS (2017) Nanostructured materials synthesis using ultrasound. Top Curr Chem 375:12
35. Hübschen G, Altpeter I, Tschuncky R, Herrmann HG (2016) Materials characterization using nondestructive evaluation (NDE) methods. Elsevier, Duxford
36. Jama M, Singh T, Gamaleldin SM, Koc M, Samara A, Isaifan RJ, Atieh MA (2016) Critical review on nanofluids: preparation, characterization, and applications. J Nanomater 6717624:1–22
37. Jeryrajkumar L, Anbarasu G, Elangovan T (2016) Effects on nano additives on performance and emission characteristics of *Calophyllim inophyllum* biodiesel. Int J Chem Tech Res 9(4):210–219
38. John S, Mathew J (2019) Determination of ferromagnetic, superparamagnetic and paramagnetic components of magnetization and the effect of magnesium substitution on structural, magnetic and hyperfine properties of zinc ferrite nanoparticles. J Magn Magn Mater 475:160–170
39. Kartick B, Srivastava SK (2011) Simple facile route for the preparation of graphite oxide and graphene. J Nanosci Nanotechnol 11:8586–8592
40. Kartick B, Srivastava SK, Srivastava I (2013) Green synthesis of graphene. J Nanosci Nanotechnol 13:4320–4324
41. Karuppuraja M, Murugesan S (2018) Template free solvothermal synthesis of single crystal magnetic Fe_3O_4 hollow spheres, their interaction with bovine serum albumin and antibacterial activities. J Saudi Chem Soc 22:569–580
42. Kegl T (2018) New nanomaterials for the adsorption of rare earth elements from aqueous solutions. University of Maribor, Faculty of Chemistry and Chemical Engineering, Maribor
43. Kegl T, Košak A, Lobnik A, Ban I (2019) Terbium ion adsorption from aqueous solution by using magnetic γ-Fe_2O_3-NH_4OH@SiO_2 nanoparticles functionalized with amino groups. Materials 12(8):1294
44. Kegl T, Ban I, Lobnik A, Košak A (2019) Synthesis and characterization of novel γ-Fe_2O_3-NH_4OH@SiO_2(APTMS) nanoparticles for dysprosium adsorption. J Hazard Mater 121632:1–29
45. Kegl T, Košak A, Lobnik A, Novak Z, Kovač Kralj A, Ban I (2020) Adsorption of rare earth metals from wastewater by nanomaterials: a review. J Hazard Mater 386:121632
46. Khorrami SA, Mahmoudzadeh G, Madani SS, Gharib F (2011) Effect of calcination temperature on the particle sizes of zinc ferrite prepared by a combination of sol-gel auto combustion and ultrasonic irradiation techniques. J Ceram Process Res 12(5):504–508
47. Koo KN, Ismail AF, Othman MHD, Rahman MA, Sheng TZ (2019) Preparation and characterization of superparamagnetic magnetite (Fe_3O_4) nanoparticles: a short review. Malays J Fundam Appl Sci 15(1):23–31

48. Kumar H, Manisha K (2018) Synthesis and characterization of silver-oxide nanoparticles by sol-gel method. Int J Adv Res Sci Eng 7(3):632–637
49. Lenin MA, Swaminathan MR, Kumaresan G (2013) Performance and emission characteristics of a DI diesel engine with a nanofuel additive. Fuel 109:362–365
50. León V, Quintana M, Herrero MA, Fierro JLG, Hoz A, Prato M, Vázquez E (2011) Few-layer graphenes from ball-milling of graphite with melanine. Chem Commun 47:10936–10938
51. Łuczak J, Paszkiewicz M, Krukowska A, Malankowska A, Zaleska-Medynska A (2016) Ionic liquids for nano- and microstructures preparation. Part 1: properties and multifunctional role. Adv Coll Interface Sci 230:13–28
52. Łuczak J, Paszkiewicz M, Krukowska A, Malankowska A, Zaleska-Medynska A (2016) Ionic liquids for nano- and microstructures preparation. Part 2: application in synthesis. Adv Coll Interface Sci 227:1–52
53. Mahmoud AED, Stolle A, Stelter M (2018) Sustainable synthesis of high-surface-area graphite oxide via dry ball milling. ACS Sustain Chem Eng 6:6358–6369
54. Mirzajanzadeh M, Tabatabaei M, Ardjmand M, Rashidi A, Ghobadian B, Barkhi M, Pazouki M (2015) A novel soluble nano-catalysts in diesel-biodiesel fuel blends to improve diesel engines performance and reduce exhaust emissions. Fuel 139:374–382
55. Mittal AK, Chisti Y, Banerjee UC (2013) Synthesis of metallic nanoparticles using plant extracts. Biotechnol Adv 31:346–356
56. Mourdikoudis S, Pallares RM, Thanh NTK (2018) Characterization techniques for nanoparticles: comparison and complementary upon studying nanoparticle properties. Nanoscale 10:12871–12934
57. Muthusamy S, Nallathambi SS, Ramasamy RK, Mohamed ST (2018) Effects of nanoparticles blended biodiesel on single cylinder CI engine. Mater Today Proc 5:6831–6838
58. Nabi G, Aain Q, Khalid NR, Tahir MB, Rafique M, Rizwan M, Hussain S, Iqbal T, Majid A (2018) A review on novel eco-friendly green approach to synthesis TiO_2 nanoparticles using different extracts. J Inorg Organomet Polym Mater 28:1552–1564
59. Nadimpalli NKV, Banyopadhyaya R, Runkana V (2018) Thermodynamic analysis of hydrothermal synthesis of nanoparticles. Fluid Phase Equilibra 456:33–45
60. Najafi A, Golestani-Fard F, Rezaie HR (2018) Sol-gel synthesis and characterization of B_4C nanopowder. Ceram Int 44:21386–21394
61. Nassar MY, Mohamed TY, Ahmed IS, Samir I (2017) MgO nanostructure via a sol-gel combustion synthesis method using different fuels: an efficient nano-adsorbent for the removal of some anionic textile dyes. J Mol Liq 225:730–740
62. Nayak S, Shet VB, Rao CV, Joshi K (2018) Performance evaluation and emission characteristics of a 4 stroke diesel engine using green synthesized silver nanoparticles blended biodiesel. Mater Today Proc 5:7889–7897
63. Nersisyan HH, Lee JH, Ding JR, Kim KS, Manukyan KV, Mukasyan AS (2017) Combustion synthesis of zero-, one-, two- and three-dimensional nanostructures: current trends and future perspectives. Prog Energy Combust Sci 63:79–118
64. Ojha PK, Karmakar S (2014) Boron for liquid fuel engines—a review on synthesis, dispersion stability in liquid fuel, and combustion aspects. Prog Aerosp Sci 100:18–45
65. Ooi JB, Ismail HM, Tan BT, Wang X (2018) Effects of graphite oxide and single-walled carbon nanotubes as diesel additives on the performance, combustion, and emission characteristics of a light-duty diesel engine. Energy 161:70–80
66. Patra JK, Baek KH (2014) Green nanobiotechnology: factors affecting synthesis and characterization techniques. J Nanomater 417305:1–12
67. Paulchamy B, Arthi G, Lignesh BD (2015) A simple approach to stepwise synthesis of graphene oxide nanomaterial. J Nanomed Nanotechnol 6(1):1000253
68. Perumal V, Ilangkumaran M (2018) The influence of copper oxide nano particle added pongamia methyl ester biodiesel on the performance, combustion and emission of a diesel engine. Fuel 232:791–802
69. Phiwdang K, Suphankij S, Mekprasart W, Pechrapa W (2013) Synthesis of CuO nanoparticles by precipitation method using different precursors. Energy Proc 34:740–745

70. Praveen A, Rao GLN, Balakrishna B (2018) Performance and emission characteristics of a diesel engine using Calophyllum Inophyllum biodiesel blends with TiO_2 nanoadditives and EGR. Egypt J Petrol 27(4):731–748
71. Qiu S, Zhao Z, Sun X (2017) Development of magnetic silica hybrid material with P507 for rare earth adsorption. J Chem Eng Data 62:469–476
72. Radhakrishnan S, Munuswamy DB, Devarajan Y, Arunkumar T, Mahalingam A (2018) Effect of nanoparticle on emission and performance characteristics of a diesel engine fueled with cashew nut shell biodiesel. Energy Sources A Recov Util Environ Eff 40(20):2485–2493
73. Rafienia M, Bigham A, Hassanzadeh-Tabrizi SA (2018) Solvothermal synthesis of magnetic spinel ferrites. J Med Sig Sens 8:108–118
74. Rafique M, Sadaf I, Rafique MS, Tahir MB (2017) A review on green synthesis of silver nanoparticles and their applications. Artifi Cells Nanomed Biotechnol 45(7):1272–1291
75. Raju VD, Kishore PS, Nanthagopal K, Ashok B (2018) An experimental study on the effect of nanoparticles with novel tamarind seed methyl ester for diesel engine applications. Energy Convers Manag 164:655–666
76. Ramachandran R, Jung D, Bernier NA, Logan JK, Waddington MA, Spokoyny AM (2018) Sonochemical synthesis of small boron oxide nanoparticles. Inorg Chem 57:8037–8041
77. Ramanan MV, Yuvarajan D (2016) Emission analysis on the influence of magnetite nanofluid on methyl ester in diesel engine. Atmosp Pollut Res 7:477–481
78. Ramasami AK, Ravishankar TN, Nagaraju G, Ramakrishnappa T, Teixeira SR, Balakrishna RG (2017) Gel-combustion-synthesized ZnO nanoparticles for visible light-assisted photocatalytic hydrogen generation. Bull Mater Sci 40(2):345–354
79. Rani S, Varma GD (2015) Superparamagnetism and metamagnetic transition in Fe_3O_4 nanoparticles synthesized via co-precipitation method at different pH. Phys B 472:66–77
80. Ranjan A, Dawn SS, Jayaprabakar J, Nirmala N, Saikiran K, Sriram SS (2018) Experimental investigation on effect of MgO nanoparticles on cold flow properties, performance, emission and combustion characteristics of waste cooking oil biodiesel. Fuel 220:780–791
81. Rao MS, Anand RB (2016) Performance and emission characteristics improvement studies on a biodiesel fuelled DICI engine using water and AlO(OH) nanoparticles. Appl Therm Eng 98:636–645
82. Runowski M (2014) Nanotechnogy—nanomaterials, nanoparticles and multifunctional core/shell type nanostructures. Chemik 9:771–779
83. Sari AM, Khazali A, Parhizgar SS (2018) Synthesis and characterization of long-CNTs by electrical arc discharge in deionized water and NaCl solution. Int Nano Lett 8:19–23
84. Sharma CK, Sharma M, Verma O, Sharma V (2015) Green synthesis of different nanoparticles and their potential applications in different fields—a critical review. Int J Pharm Bio Sci 6(3):555–567
85. Sharma R, Sharma AK, Sharma V (2015) Synthesis of carbon nanotubes by arc-discharge and chemical vapor deposition method with analysis of its morphology, dispersion and functionalization characteristics. Cogent Eng 2:1094017
86. Socrates G (2005) Infrared and Raman characteristic group frequencies. Tables and charts. Wiley, Chichester
87. Somu C, Karthi A, Singh S, Karthikeyan R, Dinesh S, Ganesh N (2017) Synthesis of various forms of carbon nanotubes by arc discharge methods—comprehensive review. Int Res J Eng Technol 4(1):344–354
88. Souri M, Hoseinpour V, Shakeri A, Ghaemi N (2018) Optimisation of green synthesis of MnO nanoparticles via utilizing response surface methodology. IET Nanobiotechnol 12(6):822–827
89. Srinidhi C, Madhusudhan A (2017) A diesel engine performance investigation fuelled with nickel oxide nano fuel-methyl ester. Int J Renew Energy Res 7(2):676–681
90. Stuart B (2004) Infrared spectroscopy: fundamentals and applications. Wiley, Sydney
91. Sundaram D, Yang V, Yetter RA (2017) Metal-based nanoenergetic materials: Synthesis, properties, and applications. Prog Energy Combust Sci 61:293–365
92. Sur UK, Ankamwar B, Karmakar S, Halder A, Das P (2018) Green synthesis of silver nanoparticles using plant extract of Shikakai and Reetha. Mater Today Proc 5:2321–2329

93. Suresh S, Pradheesh G, Ramani VA (2018) Biosynthesis and characterization of CuO, MgO and Ag_2O nanoparticles, anti-inflammatory activity and phytochemical screening of the ethanolic extract of the medicinal plant *Pavetta indica* Linn. J Pharm Phytochem 7(4):1984–1990
94. Vairamuthu G, Sundarapandian S, Kailasasanathan C, Thangagiri B (2016) Experimental investigation on the effects of cerium oxide nanoparticle on *Calophyllum inophyllum* (Punnai) biodiesel blended with diesel fuel in DI diesel engine modified by nozzle geometry. J Energy Inst 89:668–682
95. Venu H, Madhavan V (2016) Effect of Al_2O_3 nanoparticles in biodiesel-diesel-ethanol blends at various injection strategies: performance, combustion and emission characteristics. Fuel 186:176–189
96. Vinukumar K, Azhagurajan A, Vettivel SC, Vedaraman N, Lenin AH (2018) Biodiesel with nano additives from coconut shell for decreasing emissions in diesel engines. Fuel 222:180–184
97. Vinukumar K, Azhagurajan A, Vettivel SC, Vedaraman N, Lenin AH (2018) Rice husk as nanoadditive in diesel-biodiesel fuel blends used in diesel engine. J Therm Anal Calorim 131:1333–1343
98. Wang J, Li L, He Y, Song H, Chen X, Guo J (2018) The effect of thermo-oxidative ageing on crystallization, dynamic and static mechanical properties of long glass fibre-reinforced polyamide 10T composites. R Soc Open Sci 5:172029
99. Wani I, Khatoon S, Ganguly A, Ahmed J, Ganguli AK, Ahmad T (2010) Silver nanoparticles: Large scale solvothermal synthesis and optical properties. Mater Res Bull 45:1033–1038
100. Wu Q, Xie X, Wang Y, Roskilly T (2018) Effect of carbon coated aluminum nanoparticles as additive to biodiesel-diesel blends on performance and emission characteristics of diesel engine. Appl Energy 221:597–604
101. Xu R, Pang W, Huo Q (2011) Modern inorganic synthetic chemistry. Elsevier, Amsterdam
102. Xu H, Zeiger BW, Suslick KS (2013) Sonochemical synthesis of nanomaterials. Chem Soc Rev 42:2555–2567
103. Yadav TP, Yadav RM, Singh DP (2012) Mechanical milling: a top down approach for the synthesis of nanomaterials and nanocomposites. Nanosci Nanotechnol 2(3):22–48
104. Yarestani M, Khalaji AD, Rohani A, Das D (2014) Hydrothermal synthesis of cobalt oxide nanoparticles: Its optical and magnetic properties. J Sci Islamic Repub Iran 25(4):339–343
105. Zaaba NI, Foo KL, Hashim U, Tan SJ, Liu WW, Voon CH (2017) Synthesis of graphene oxide using modified hummers method: solvent influence. Proc Eng 184:469–477
106. Zhang H, Wu Q, Lin J, Chen J, Xu Z (2010) Thermal conductivity of polyethylene glycol nanofluids containing carbon coated metal nanoparticles. J Appl Phys 108(12):124304
107. Zimur SD, Kamble PD, Gaikwad PV, Mali AV (2018) Effect of Sn substitution on structural, electrical and magnetic properties of copper ferrites synthesized by sol-gel auto combustion method. Int J Res Anal Rev 5(4):139–148

Chapter 4
Nanofuels

By adding NMs to various diesel engine base fuels, one gets NMs-enriched fuels, or nanofuels in short. This addition may influence significantly the properties and behavior of the fuel due to unique properties of NMs [31, 44, 94]. This chapter discusses the preparation and properties of such nanofuels.

The considered NMs are those listed in Tables 3.3–3.11 (Sect. 3.3), either used individually or mixed in various combinations. The base fuels (Sect. 2.3), used with these additives, are:

- neat fuels like D100, *ci*B100, *cns*B100, *j*B100, *m*B100, *mo*B100, *n*B100, *p*B100, *pp*B100, *r*B100, *rb*B100, *s*B100, *wco*B100, GTL100,
- blends of mineral diesel and biodiesel like D/*aa*B20, D/*bba*B20, D/*ci*B20, D/*ci*B50, D/*co*B20, D/*cs*B10, D/*cs*B20, D/*gs*B20, D/*h*B20, D/*j*B20, D/*jj*B20, D/*l*B20, D/*lg*B25, D/*m*B20, D/*m*B25, D/*m*B50, D/*me*B20, D/*p*B20, D/*pl*B20, D/*po*B20, D/*pp*B20, D/*pp*B25, D/*s*B50, D/*sg*B20, D/*ts*B20, D/*wco*B10, D/*wco*B20, D/*wco*B2, D/*wfo*B20, D/*zj*B25,
- blends of biodiesel and oil like *j*B/*tp*O10,
- blends of mineral diesel and ethanol like D/E4 and D/E35,
- blends of mineral diesel, biodiesel, and butanol or pentanol like D/wcoB20/Bu10 and D/*c*B25/P20,
- blends of biodiesel and butanol like *j*B/Bu40,
- blends of mineral diesel, biodiesel, and ethanol like D/*j*B20/E10, D/*jj*B10/E20, D/*p*B10/E4, D/*p*B30/E30, and D/*s*B15/E5,
- water-in-diesel emulsions, like D/W10 and D/W5,
- water-in-oil emulsions like *lg*O/W5 and *p*O/W5,
- water-in-biodiesel emulsions like *j*B/W5, *j*B/W10, and *no*B/W15, and
- water-in-diesel/biodiesel emulsions like D/*r*B5/W4, D/*wco*B15/W1, and D/*wco*B5/W3.

The most important properties of these fuels, both pure and enriched with NMs, are given in tabular form for each group of NMs in Sect. 4.2.

T. Kegl et al., *Nanomaterials for Environmental Application*, Green Energy and Technology, https://doi.org/10.1007/978-3-030-54708-0_4

4.1 Nanofuel Preparation

Nanofuels are most frequently synthesized by using either single-step or two-step methods [5, 99].

Single-step methods are vapor deposition, laser ablation, and submerged arc methods, where the preparation and suspension of NMs into the base fuel are carried out synchronously [5]. In this way, the NMs storage, drying, and transportation are avoided, and therefore the agglomeration of NMs is minimized, which in turn results in fine dispersion of NMs in the base fuel [5].

By **two-step methods,** the NMs are synthesized in a dedicated process and stored for later use. The prepared nanopowder is then dispersed into the base fuel as needed, either during the

- production process (PP) of the base fuel (e.g., transesterification process for biodiesel fuels) [73], or
- post-production process (P-PP) of the base fuel (e.g., post-transesterification process for biodiesel fuels).

The dispersion process is done by engaging either ultrasonication methods, magnetic stirring (homogenization, high shear mixing), pH adjustment, or any combination of these [5]. Ultrasonication methods, probe-type sonication, and ultrasonic bath are processes in which sound waves are used to agitate the NMs in a solution. NMs are dispersed in a fluid by means of an ultrasonicator, which transfers ultrasonic energy to the particles for better dispersion [68]. Magnetic stirring is used to disperse NMs with low concentration. In this technique, the stir bar, which is submerged into the solution, is rotated very fast under the action of a rotating magnetic field, created by stationary electromagnets or a set of rotating electromagnets underneath the vessel in a stirring apparatus. This technique is usually used for prehomogenization of nanofuels [5]. The method of pH adjustment is used to set up an appropriate dispersion environment by adjusting the pH-value of the nanofuel. In practice, to obtain nanofuels with the desired dispersion efficiency and stability, adequate combinations of ultrasonication, magnetic stirring, and pH adjustment methods are used very frequently. Two-step methods are the most economical and widely used methods for preparing nanofuels [81]. The main problem related to these methods is the agglomeration of NMs and insufficient stability of nanofuels. To avoid or reduce NMs agglomeration and to improve the stability, surfactants are often engaged [5]. A surfactant reduces the surface tension between molecules of various fuel components [5, 99, 104]. Efficient and frequently used surfactants are:

- 2-ethylhexyl nitrate (2-EN),
- ammonium polymethacrylate (Darvan-C),
- cetyl trimethyl ammonium bromide (CTAB),
- sorbitan oleate,
- sorbitan monooleate (Span 80^{TM}),
- 2-[2-[3,4-bis(2-hydroxyethoxy) oxolan-2-yl]-2-(2-hydroxy ethoxy) ethoxy]ethyl octadec-9-enoate (Tween 80^{TM}),

Table 4.1 Preparation of nanofuels by using NMs based on silver

Nanoadditive	NMs: size dosage	Base fuel	Surfactant (solvent)	Two-step synthesis method of nanofuel	Stability aspects	References
Ag NPs	30–50 nm 40 ppm	D100	Span 80™	P-PP: homogenization, ultrasonication (10 min)	Stable	[87]
	80–90 nm 50 ppm	*pp*B100	None	P-PP: ultrasonication (40 kHz, 30 min)	Stable	[57]
	50 nm 120 ppm	*wco*B100	None	P-PP: ultrasonication (bath)	Stable	[54]
Ag_2O NPs	<100 nm 10 ppm	*n*B100	None	P-PP: ultrasonication (45 kHz, 220 W)	72 h	[17]
	<100 nm 10 ppm	*p*B100	None	P-PP: ultrasonication (60 kHz, 440 W)	Stable	[18]

- oleic acid,
- citric acid,
- mixture of Span 80™ and Tween-80 ™,
- mixture of Span 80™ and EtOH,
- isopropanol,
- trimethyl ammonium hydroxide (TMAOH), and
- sodium dodecyl sulfate (SDS).

The surfactants, preparation, and stability aspects of nanofuels by using metallic and nonmetallic NMs are collected in Tables 4.1, 4.2, 4.3, 4.4, 4.5, 4.6, 4.7, 4.8, 4.9, 4.10, 4.11, 4.12, and 4.13.

NMs addition to the base fuel influences the nanofuel stability, which depends on the base fuel, NMs type, NMs size, NMs dosage, engaged preparation method, and used surfactants. Long-term stability of nanofuels was reported only for a few tested nanofuels, Tables 4.1, 4.2, 4.3, 4.4, 4.5, 4.6, 4.7, 4.8, 4.9, 4.10, 4.11, 4.12, and 4.13 [12, 20, 45, 64, 75, 88]. From what was reported so far, nanofuel stability in the long term can be reached in the following ways:

- by addition of suitable surfactants such as 2-ethylhexyl nitrate for Al_2O_3-D100 [12], citric acid for γ-Fe_2O_3-D/*pp*B20 [45], sorbitan oleate for CeO_2-D100 [88], TMAOH for Fe_3O_4-*rb*B100 [75], and SDS for G-D/*sg*B20 [64],
- by appropriate preparation method without adding any surfactant; this was confirmed by mwC-*j*B/Bu40 and G-*j*B/Bu40 nanofuels prepared with two ultrasonications at different conditions [20].

Table 4.2 Preparation of nanofuels by using NMs based on aluminum

Nanoadditive	NMs: size dosage	Base fuel	Surfactant (solvent)	Two-step synthesis method of nanofuel	Stability aspects	References
Al_2O_3 NPs	50 nm 10 wt%	D100	2-EN (dH_2O)	Post-PP: ultrasonication (shaker: 1 h)	Long time	[12]
	20–30 nm 100 ppm	D100	None	P-PP: ultrasonication (400 W, 24 kHz, 30 min)	200 h	[13]
	27–43 nm 100 ppm	D100	Darvan-C	P-PP: homogenization, mixing ultrasonication (40 Hz, 1 h)	Stable	[28, 32]
	≤50 nm 100 ppm	D100	None	P-PP: ultrasonication (probe-type)	Week	[38]
	5–150 nm 0.5 wt%	D100	Span 80™	P-PP: ultrasonication (750 W, 20 kHz, 15 min)	Stable	[50]
	<100 nm 0.006 wt%	D100	None	Post-PP: ultrasonication	–	[32, 53]
	45 nm 150 ppm	D100	None	P-PP: ultrasonication (mixing: 1 h, cleaner bath: 6 h)	–	[56]
	34 nm 50 ppm	D/W10	Span 80™	P-PP: mixing	Stable	[29]
	30–50 nm 40 ppm	D/*ci*B20	CTAB	P-PP: ultrasonication (50 Hz, 120 W, 45 min)	–	[6]
	50 nm 100 ppm	D/*gs*B20	None	P-PP: ultrasonication (40 kHz, 120 W, 1 h)	–	[41]
	<50 nm 0.5 g/L	D/*j*B20	None	P-PP: ultrasonication (vibrator)	–	[65]
	20–50 nm 30 mg/L	D/*jj*B20	None	P-PP: ultrasonication (24 kHz, 30 min)	Week	[24]

(continued)

Table 4.2 (continued)

Nanoadditive	NMs: size dosage	Base fuel	Surfactant (solvent)	Two-step synthesis method of nanofuel	Stability aspects	References
	32–48 nm 100 ppm	D/*m*B20	CTAB	P-PP: ultrasonication (shaker: 30–45 min)	–	[1]
	30–50 nm 100 ppm	D/*m*B25	Tween 80™	P-PP: ultrasonication (10 min)	–	[79]
	20–50 nm 30 mg/L	D/*pl*B20	Yes	P-PP: ultrasonication (24 kHz, 30 min)	Stable	[77]
	<100 nm 100 ppm	D/*pp*B25	CTAB (EtOH)	P-PP: stirring, ultrasonication (24 kHz, 30 min)	Stable	[93]
	20 nm 60 ppm	D/*ts*B20	None	P-PP: stirring (magnetic stirrer) and sonication (ultrasonicator: 120 W, 40 kHz)	Stable	[74]
	50 nm 90 ppm	D/*wco*B10	None	P-PP: ultrasonication (15 min)	3 h	[32]
	<100 nm 25 ppm	D/*zj*B25	None	P-PP: homogenization, ultrasonication (30 min)	–	[2]
	<100 nm 10 wt%	*cns*B100	None (EtOH)	PP: ultrasonication (45 kHz, 220 W)	–	[73]
	30 nm 50 ppm	D/E35	None	P-PP: ultrasonication (40 kHz)	–	[78]
	20 nm 50 mg/L	D/*j*B20/E10	None	P-PP: stirring (120 min), ultrasonication (15 min)	–	[100]

(continued)

Table 4.2 (continued)

Nanoadditive	NMs: size dosage	Base fuel	Surfactant (solvent)	Two-step synthesis method of nanofuel	Stability aspects	References
	20–50 nm 75 mg/L	D/*jj*B10/E20	None	P-PP: homogenization (23,000 rpm), ultrasonication (bath: 24 kHz, 30 min)	Stable	[58]
	50 nm 100 mg/L	D/*s*B15/E5	Isopropanol	P-PP: mixing, ultrasonication	Stable	[91]
	27–43 nm 2 wt%	GTL100	Span 80™	P-PP: ultrasonication (probe-type: 1 h, temperature controlled)	2 h	[39]
AlOOH NPs	15 nm 200 ppm	D/*co*B20	Span 80™ (dH_2O)	P-PP: ultrasonication (50–60 kHz, 45 min)	5 days	[72]
	<20 nm 100 ppm	*j*B/W10	Span 80™ Tween 80™	P-PP: ultrasonication (120 W, 40 kHz, 30 min)	>week	[81]

Furthermore, NMs size influences the nanofuel stability to a great extent. Rao and Dash [82] investigated the influence of Al_2O_3 NPs size on the Al_2O_3-*j*B100 nanofuel stability, by using Span^{80} and Tween^{80} as stabilizing surfactant. They found out that by reducing the size of the used particles by a factor of two, fuel stability improved from half an hour up to one year. The possible reasons are the higher surface area-to-volume ratio and large liquid/solid interface of smaller particles [89]. Increased stability was achieved also by reducing the CeO_2 NPs size from 50 to 10–30 nm [59, 88] and the TiO_2 NPs size from 45 to 10–20 nm [14, 56] when added to D100. This trend is also confirmed by the fact that all nanofuels with ascertained long-term stability, given in Tables 4.1, 4.2, 4.3, 4.4, 4.5, 4.6, 4.7, 4.8, 4.9, 4.10, 4.11, 4.12, and 4.13, where prepared with NMs exhibiting relative low sizes. It can be concluded that for long-term stability the NMs size, surfactants, and preparation method conditions (e.g., sonication time, wave, and pulse mode; mixing time and speed, etc.) have to be optimized.

As can be seen from the synthesis method column of Tables 4.1, 4.2, 4.3, 4.4, 4.5, 4.6, 4.7, 4.8, 4.9, 4.10, 4.11, 4.12, and 4.13, to prepare a stable and homogeneous nanofuel, various synthesized NMs might need to be first dispersed in a solvent; either in dH_2O [8, 9, 11, 12, 19, 25, 45, 55, 104], or EtOH [73, 76, 93, 103], or in butanol [20, 23]. In these cases, the so-called *nanofluid* (this term will be used throughout

Table 4.3 Preparation of nanofuels by using NMs based on cerium

Nanoadditive	NMs: size dosage	Base fuel	Surfactant (solvent)	Two-step synthesis method of nanofuel	Stability aspects	References
CeO_2 NPs	15–30 nm 25 mg/L	D100	None	P-PP: mixing (90 min), ultrasonication (shaker)	48 h	[26]
	10–30 nm 40 ppm	D100	Sorbitan oleate	P-PP: ultrasonication (15 min, ice bath)	Long term	[88]
	50 nm 25 ppm	D100	Sorbitan oleate	P-PP: ultrasonication (40 kHz, 120 W, 30 min)	Stable	[59]
	16 nm 30 ppm	*lg*O/W5	Span 80™	P-PP: mechanical stirring, ultrasonication	4 days	[7]
	25 nm 40 ppm	D/*ci*B50	–	P-PP: -	–	[97]
	<100 nm 50 ppm	D/*cs*B10	Oleic acid	P-PP: ultrasonication	Stable	[66]
	<100 nm 40 mg/L	D/*cs*B20	– ; (EtOH)	P-PP: ultrasonication (ultrasonicator)	Stable	[76]
	<100 nm 20 ppm	D/*lg*B25	None	P-PP: ultrasonication	–	[19]
	<25 nm 150 ppm	D/*m*B20	CTAB	P-PP: ultrasonication (shaker: 30–45 min)	–	[46]
	<100 nm 90 ppm	D/*r*B5/W4	–	P-PP: –	–	[37]
	<100 nm 90 ppm	D/*wco*B5/W3	Span 80™+ Tween 80™	P-PP: homogenization (15 min)	Stable	[44]
	15–30 nm 90 ppm	D/*wco*B5/W3	Span 80™+ Tween 80™	P-PP: homogenization (room temperature, 15 min)	–	[3, 4]

(continued)

Table 4.3 (continued)

Nanoadditive	NMs: size dosage	Base fuel	Surfactant (solvent)	Two-step synthesis method of nanofuel	Stability aspects	References
	25 nm 40 ppm	*ci*B100	–	P-PP: -	–	[6, 97]
	<100 nm 40 ppm	*j*B100	None	P-PP: mixing, ultrasonication (30 min)	Used after synthesis	[65, 85]
	20–30 nm 100 ppm	*j*B/*tp*O10	None	P-PP: ultrasonication (60 min)	–	[92]
	15 nm 30 ppm	*no*B/W15	Span 80™+ EtOH (dH_2O)	P-PP: ultrasonication (20 kHz, 30 min)	62 h	[19]

Table 4.4 Preparation of nanofuels by using NMs based on cobalt

Nanoadditive	NMs: size dosage	Base fuel	Surfactant (solvent)	Two-step synthesis method of nanofuel	Stability aspects	References
Co_3O_4 NPs	<100 nm 100 ppm	D/*gs*B20	None	P-PP: ultrasonication	–	[42]
	30 nm 50 ppm	D/*wfo*B20	None	P-PP: ultrasonication (bath: 30 min)	Stable	[49]
	<100 nm 150 mg/L	*ci*B100	CTAB	P-PP: ultrasonication	Highly stable	[35]
	<100 nm 100 ppm	*j*B100	CTAB	P-PP: homogenization, mixing, ultrasonication	Stable	[65, 83]

this paper) is obtained. This fluid replaces the usual NMs in solid form (powder) in the nanofuel preparation process. In dependence on the base fuel, surfactant, and nanofuel preparation method, the finally obtained nanofuel droplets can contain either encapsulated NMs (NMs powder additive) or encapsulated NMs into dH_2O or alcohol (NMs nanofluid additive), Fig. 4.1.

Under the bottom line, the main challenge of each preparation method for a nanofuel is to make a homogeneous and long-term stable fuel with negligible agglomeration. Of course, this has to be achieved without affecting negatively the chemical and thermo-physical properties of the nanofuel.

Table 4.5 Preparation of nanofuels by using NMs based on copper

Nanoadditive	NMs: size dosage	Base fuel	Surfactant (solvent)	Two-step synthesis method of nanofuel	Stability aspects	References
CuO NPs	30–50 nm 50 ppm	D100	Darvan-C	P-PP: homogenization, mixing, ultrasonication (1 h, 40 Hz)	Stable	[28]
	<100 nm 0.006 wt%	D100	None	P-PP: ultrasonication	–	[53]
	<100 nm 0.5 wt%	D100	Span 80™	P-PP: mixing, ultrasonication (1 h, 40 Hz)	Stable	[84]
	32 nm 50 ppm	D/W10	Span 80™	P-PP: mixing	Stable	[29]
	<100 nm 80 ppm	D/*l*B20	None	P-PP: ultrasonication (20 kHz, 30 min)	–	[34]
	<50 nm 50 mg/L	D/*pp*B20	None	P-PP: ultrasonication (bath)	Used after synthesis	[16]
	50 nm 50 ppm	D/*pp*B20	Span 80™	P-PP: ultrasonication (45 min)	Stable	[67]

4.2 Nanofuel Properties

Tables 4.14, 4.15, 4.16, 4.17, 4.18, 4.19, 4.20, 4.21, 4.22, 4.23, 4.24, and 4.25 offer an overview of the following most important fuel properties: fuel density (ρ_f), kinematic viscosity (ν_f), flash point (T_{fp}), cetane number (CN), and calorific value (H_f). For the base fuels, absolute numbers are listed as reported in the corresponding literature. For nanofuels, a relative variation of the corresponding property is provided. This data is obtained as a percentage of nanofuel property variation divided by the corresponding property value of the base fuel. It has to be pointed out that the properties of the same base fuel type may vary because they are listed as reported by the authors and no attempt is made to investigate deeply the reasons for these variations. Namely, the objective here is to focus on the changes in fuel properties, caused by adding various NMs.

Table 4.6 Preparation of nanofuels by using NMs based on iron

Nanoadditive	NMs: size dosage	Base fuel	Surfactant (solvent)	Two-step synthesis method of nanofuel	Stability aspects	References
γ-Fe_2O_3 NPs	35 nm 100 ppm	D100	None	P-PP: ultrasonication (probe-type)	12 h	[38]
	30–43 nm 0.5 wt%	D100	Span 80™	P-PP: ultrasonication (750 W, 20 kHz, 15 min)	Stable	[50]
	30 nm 1 wt%	D/*pp*B20	Citric acid (dH_2O)	P-PP: mixing	Long term	[45]
	<100 nm 100 ppm	D/*pp*B20	None	P-PP: mixing, ultrasonication (40 kHz, 120 W, 1 h)	–	[52]
Fe_3O_4 NPs	10 nm 1 wt%	*rb*B100	TMAOH (dH_2O)	P-PP: mixing, ultrasonication (1 h)	Long term	[75]

Table 4.7 Preparation of nanofuels by NMs based on magnesium

Nanoadditive	NMs: size dosage	Base fuel	Surfactant (solvent)	Two-step synthesis method of nanofuel	Stability aspects	References
MgO NPs	33 nm 50 ppm	D/W10	Span 80™	P-PP: mixing	Stable	[29]
	20–30 nm 30 ppm	D/*wco*B20	None	P-PP: stirring (500 rpm, 30 min), ultrasonication (10 min)	Stable	[80]
	10–30 nm 50 ppm	*r*B100	None	P-PP: mixing, ultrasonication (20 kHz, 750 W)	Used after synthesis	[61]
	20–30 nm 30 ppm	*wco*B100	None	P-PP: stirring (500 rpm, 30 min), ultrasonication (10 min)	Stable	[80]

Table 4.8 Preparation of nanofuels by NMs based on titanium

Nanoadditive	NMs: size dosage	Base fuel	Surfactant (solvent)	Two-step synthesis method of nanofuel	Stability aspects	References
TiO_2 NPs	45 nm 150 ppm	D100	None	P-PP: mixing (6 h), ultrasonication (bath: 40 kHz, 720 W, 20 min)	Used after synthesis	[56]
	10–20 nm 80 mg/L	D100	None	P-PP: ultrasonication (probe-type: 20 min)	Stable	[14, 15]
	10–50 nm 40 ppm	D/W5	Span 80™+ Tween 80™	P-PP: magnetic stirring (30 min), ultrasonication (15 min)	Stable	[96]
	10–20 nm 25 ppm	D/*me*B20	CTAB	P-PP: ultrasonication (shaker)	Stable	[47]
	10–50 nm 40 ppm	*p*O/W5	Span 80™+ Tween 80™	P-PP: magnetic stirring (30 min), ultrasonication (15 min)	Stable	[96]
	<100 nm 40 ppm	D/*ci*B20	CTAB	P-PP: –	Stable	[71]
	<100 nm 0.01 wt%	D/*wco*B20	None	P-PP: ultrasonication (homogenizator: 10 kHz, 1 h)	–	[60]
	<100 nm 150 mg/L	*ci*B100	CTAB	P-PP: magnetic stirring, ultrasonication	Highly stable	[35]
	20 nm 100 ppm	*ci*B100	Span 80™ (dH_2O)	P-PP: stirring (1500 rpm), ultrasonication (50–60 kHz, 30 min)	Stable	[10]
	20 nm 100 ppm	*ci*B100	None (dH_2O)	P-PP: ultrasonication (50–60 kHz, 30 min)	–	[55]

(continued)

Table 4.8 (continued)

Nanoadditive	NMs: size dosage	Base fuel	Surfactant (solvent)	Two-step synthesis method of nanofuel	Stability aspects	References
	60 nm 200 ppm	*m*B100	None	P-PP: ultrasonication (60–100 kHz)	–	[62]
	50 nm 200 ppm	*mo*B100	Span 80™ (dH_2O)	P-PP: ultrasonication (60–100 kHz)	7 days	[104]
	<100 nm 0.01 wt%	D/*wco*B20/Bu10	None	P-PP: ultrasonication (10 kHz, 1 h)	–	[60]

Table 4.9 Preparation of nanofuels by NMs based on zinc

Nanoadditive	NMs: size dosage	Base fuel	Surfactant (solvent)	Two-step synthesis method of nanofuel	Stability aspects	References
ZnO NPs	67 nm 100 ppm	D100	None	P-PP: ultrasonication (probe-type)	Week	[38]
	<100 nm 100 ppm	D100	–	P-PP: ultrasonication (20 kHz, 30 min)	–	[90]
	34 nm 50 ppm	D/W10	Span 80™	P-PP: mixing	Stable	[29]
	<100 nm 100 ppm	D/*m*B20	–	P-PP: ultrasonication (20 kHz, 30 min)	–	[90]
	<100 nm 100 ppm	D/*m*B50	–	P-PP: ultrasonication (20 kHz, 30 min)	–	[90]
	35 nm 100 ppm	*ci*B100	None (dH_2O)	P-PP: ultrasonication (50–60 kHz, 30 min)	–	[55]
	50 nm 100 ppm	*ci*B100	Span 80™ (dH_2O)	P-PP: ultrasonication (50–60 kHz, 30 min), stirring (mechanical: 1500 rpm)	Stable	[9]
	30 nm 250 ppm	D/*p*B30/E30	None	P-PP: ultrasonication (probe-type: 50–60 kHz, 120 W)	Stable	[68]

4.2.1 Nanofuels with NMs Based on Silver

The effects adding either Ag NPs or Ag_2O NPs to various base fuels (D100, *n*B100, *p*B100, *pp*B100, and *wco*B100) are rather similar, Table 4.14. In general, Ag and Ag_2O NPs increase the density, kinematic viscosity, cetane number, and calorific value, while they reduce the flash point [3, 17, 18, 54, 57, 87, 93].

Table 4.10 Preparation of nanofuels by NMs with C NTs

Nanoadditive	NMs: size dosage	Base fuel	Surfactant (solvent)	Two-step synthesis method of nanofuel	Stability aspects	References
C NTs	40–60 nm 100 ppm	D100	None	P-PP: ultrasonication (emulsifier: 400 W, 24 kHz, 30 min)	10 h	[13]
	<100 nm 100 ppm	D100	None	P-PP: ultrasonication (200–400 W, 50–60 Hz, 30 min)	–	[98]
	<100 nm 50 ppm	D/*h*B20	–	P-PP: ultrasonication, mechanical homogenization	–	[86]
	<100 nm 100 ppm	D/*s*B50	None	P-PP: ultrasonication (200–400 W, 50–60 Hz, 30 min)	–	[98]
	33 nm 60 ppm	D/*ts*B20	None	P-PP: mixing (15 min), ultrasonication (homogenizer)	Stable	[74]
	4–8 nm 60 ppm	D/*wco*B2	None	P-PP: stirring (magnetic stirrer) and ultrasonication (120 W, 40 kHz)	Stable	[30]
	<100 nm 100 ppm	*s*B100	None	P-PP: ultrasonication (200–400 W, 50–60 Hz, 30 min)	–	[98]
	<200 nm 100 ppm	*j*B/*tp*O10	None	P-PP: ultrasonication (1 h)	Stable	[92]
	16 nm 100 ppm	*j*B/W5	Span 80™+ Tween 80™ (dH_2O)	P-PP: mechanical agitator, ultrasonication, reactor vessel (3000 rpm)	–	[11]

(continued)

Table 4.10 (continued)

Nanoadditive	NMs: size dosage	Base fuel	Surfactant (solvent)	Two-step synthesis method of nanofuel	Stability aspects	References
	5–20 nm 100 ppm	*wco*B100	none	P-PP: ultrasonication (400 W, 24 kHz)	–	[3, 54]

4.2.2 Nanofuels with NMs Based on Aluminum

The addition of Al_2O_3 NPs to various tested base fuels typically increases the density and cetane number of the fuel, Table 4.15. The calorific value of the Al_2O_3-enriched fuels is practically always higher, except in case of the D/*wco*B10 and D/*j*B20/E10 base fuels [32, 100]. The kinematic viscosity mostly decreases and the flash point increases by addition of Al_2O_3 NPs to D100, while in the case of addition of Al-based NMs to the blends of diesel, biodiesel, and EtOH, this effect is not so clear [1, 2, 6, 28, 29, 39, 41, 56, 73, 74, 77, 91, 93].

4.2.3 Nanofuels with NMs Based on Cerium

CeO_2 NPs addition to various base fuels typically increases the density and kinematic viscosity, while the effect on other fuel properties depends more or less on size and dosage of CeO_2 NPs as well as on the type of the base fuel, Table 4.16 [7, 19, 26, 37, 44, 65, 66, 85, 92].

4.2.4 Nanofuels with NMs Based on Cobalt

The addition of Co_3O_4 NPs in general increases the density, viscosity, flash point, cetane number, and calorific value of the fuel, Table 4.17 [42, 49].

4.2.5 Nanofuels with NMs Based on Copper

CuO NPs addition to all tested base fuels increases the density, cetane number, and calorific value of the fuel, Table 4.18. The viscosity of nanofuels, prepared from mineral diesel with various dosages and sizes of CuO NPs, is lower than those of the base fuel [16, 28, 29, 53, 67, 84].

Table 4.11 Preparation of nanofuels by NMs based on carbon, graphite, and graphene

Nanoadditive	NMs: size dosage	Base fuel	Surfactant (solvent)	Two-step synthesis method of nanofuel	Stability aspects	References
swC NTs	1–2 nm 25 ppm	D100	Sorbitan oleate	P-PP: ultrasonication (40 kHz, 120 W, 30 min)	Stable	[59]
mwC NTs	10–15 nm 30 mg/L	D/*jj*B20	None	P-PP: ultrasonication (24 kHz, 30 min)	–	[21]
	10–20 nm 90 ppm	D/*wco*B10	None	P-PP: mixing (15 min), ultrasonication	3 h	[33]
	5 nm 30 ppm	D/*wco*B20	None	P-PP: ultrasonication	–	[27]
	10–15 nm 50 mg/L	*j*B/Bu40	None (Bu)	P-PP: ultrasonication (bath: 40–59 kHz, 250 W, 15 min); ultrasonication (bath: 24 kHz, 15 min)	Long-term stable	[20]
G'O NPs	1–2 nm 25 ppm	D100	Sorbitan oleate	P-PP: ultrasonication (40 kHz, 120 W, 30 min)	60 min	[33, 59]
Cqd NPs	1–3 nm 60 ppm	D/*wco*B15/W1	None (dH_2O)	P-PP: mixing, stirring, ultrasonication (bath: 5 min)	6 months	[25]
G NPTs	6–8 nm 100 mg/L	D/*j*B20	None	P-PP: homogenization (23,000 rpm), ultrasonication (bath: 24 kHz, 30 min)	5 days	[22]
	22–26 nm 40 ppm	D/*sg*B20	SDS	P-PP: homogenization	Long term	[64]

(continued)

Table 4.11 (continued)

Nanoadditive	NMs: size dosage	Base fuel	Surfactant (solvent)	Two-step synthesis method of nanofuel	Stability aspects	References
	6–8 nm 50 mg/L	*j*B/Bu40	None	P-PP: ultrasonication (bath: 15 min); ultrasonication (bath: 24 kHz, 15 min)	Long-term stable	[20]
GO NPs	6 nm 50 mg/L	*j*B100	None (Bu)	P-PP: ultrasonication (bath: 15 min); ultrasonication (bath: 24 kHz, 15 min)	3 h	[23]
	1–2 nm 90 ppm	D/*aa*B20	None	P-PP: ultrasonication (20 min)	4 h	[31]

Table 4.12 Preparation of nanofuels with organic NMs

Nanoadditive	NMs: size dosage	Base fuel	Surfactant (solvent)	Two-step synthesis method of nanofuel	Stability aspects	References
CS NPs	20 nm 0.1 wt%	D/*pp*B20	–	P-PP: mixing (atmospheric pressure, 1000 rpm)	–	[101]
RH NPs	<100 nm 0.1 wt%	D/*po*B20	None	P-PP: ultrasonication (bath)	72 h	[63]
	20 nm 0.1 wt%	D/*pp*B20	None	P-PP: ultrasonication (30 min)	30 days	[102]

4.2.6 Nanofuels with NMs Based on Iron

The addition of γ-Fe_2O_3 NPs and Fe_3O_4 NPs to various tested base fuels increases all considered fuel properties, Table 4.19 [38, 45, 52, 75].

Table 4.13 Preparation of nanofuels by NMs based on others elements

Nanoadditive	NMs: size dosage	Base fuel	Surfactant (solvent)	Two-step synthesis method of nanofuel	Stability aspects	References
B_2O_3 NPs	80–100 nm 0.5 wt%	D100	Span 80™	P-PP: ultrasonication (750 W, 20 kHz, 15 min)	Stable	[50]
MnO NPs	33 nm 50 ppm	D/W10	Span 80™	P-PP: mixing	Stable	[29]
Mn_2O_3 NPs	30 nm 50 ppm	D/*wfo*B20	None	P-PP: ultrasonication (bath: 30 min)	Stable	[49]
NiO NPs	<20 nm 40 ppm	D/*p*B20	Isopropanol	P-PP: stirring (1 h)	Stable	[95]
SiO_2 NPs	20–30 nm 100 ppm	D100	None	P-PP: ultrasonication (400 W, 24 kHz, 30 min)	200 h	[13]
	10–30 nm 50 ppm	*r*B100	None	P-PP: mixing, ultrasonication (20 kHz, 750 W)	Used after synthesis	[61]
Al_2O_3 and CeO_2 NPs	51 and 32 nm 30 ppm	D/*j*B20	None	P-PP: ultrasonication (45 min)	12 h	[69, 70]
	51 and 32 nm 30 ppm	*j*B100	None	P-PP: ultrasonication (45 min)	12 h	[69, 70]
SiO_2 and TiO_2 NPs	50 nm 100 ppm	D/*bba*B20	None	P-PP: ultrasonication (40 kHz, 120 W, 1 h)	–	[40]
Al@C NPs	60–160 nm 30 ppm	D/*p*B10/E4	None (EtOH)	P-PP: ultrasonication (bath: 30 min)	15 days	[103]
	50–60 nm 30 ppm	D/E4	None (EtOH)	P-PP: ultrasonication (bath: 30 min)	2 weeks	[36]
CeO_2/mwC (DMF) NTs	40–50 nm 90 ppm	D/*wco*B20	None	P-PP: ultrasonication (bath: 40 Hz, 30 min)	Highly stable	[51]

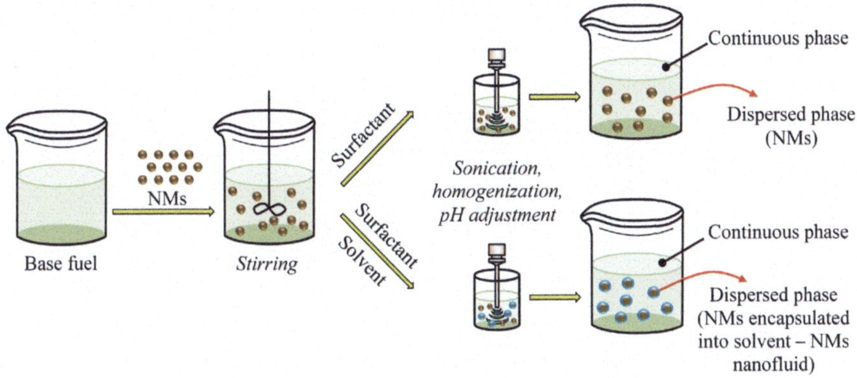

Fig. 4.1 Nanofuels preparation options

4.2.7 Nanofuels with NMs Based on Magnesium

The addition of MgO NPs to various base fuels increases all considered fuel properties, Table 4.20 [29, 43, 61, 80].

4.2.8 Nanofuels with NMs Based on Titanium

TiO_2 NPs addition to various base fuels increases the fuel density and viscosity, Table 4.21. In some cases, the nanofuel exhibits smaller calorific values, smaller cetane numbers, and smaller flash points, when compared to the base fuel [10, 14, 47, 55, 56, 60, 62, 71, 96, 104].

4.2.9 Nanofuels with NMs Based on Zinc

ZnO NPs addition to various base fuels increases all listed fuel properties with the exceptions of smaller calorific values, smaller flash points, and smaller viscosities in some cases, Table 4.22 [9, 29, 38, 55, 68].

4.2.10 Nanofuels with NMs with C NTs

Addition of C NTs to tested base fuels increases the fuel density, Table 4.23. The effects of C NTs addition to various base fuels on viscosity, flash point, cetane number, and calorific value vary in dependence on the size and dosage of NTs [3, 11, 13, 30, 54, 86, 92, 98].

Table 4.14 Properties of nanofuels with NMs based on silver

Nanoadditive	Nanoadditive: size dosage	Base fuel	Base fuel properties					Nanofuel properties with respect to base fuel					References
			ρ_f	ν_f	T_{fp}	CN	H_f	ρ_f	ν_f	T_{fp}	CN	H_f	
			$\left(\frac{kg}{m^3}\right)$	$\left(\frac{mm^2}{s}\right)$	(°C)	(/)	$\left(\frac{MJ}{kg}\right)$	Maximum % increase (↑) or decrease (↓)					
Ag NPs	30–50 nm 40 ppm	D100	825–855	1.3–4.1	60–80	48–57	42–45	–	↑	–	–	↑	[48, 87]
	50 nm 120 ppm	*wco*B100	870–890	4.6–4.8	170–180	57–63	37–39	–	–	–	↑ 2.8	–	[3, 54]
Ag_2O NPs	<100 nm 10 ppm	*n*B100	844	4.28	140	55	37.5	↑ 2.1	↑ 1.6	↓ 2.9	↑ 3.6	↑ 1.6	[17]
	<100 nm 10 ppm	*p*B100	801	3.85	146	53	39.4	↑ 2.1	↑ 3.1	↓ 4.8	↑ 7.5	↑ 1.6	[18]

Table 4.15 Properties of nanofuels with NMs based on aluminum

Nanoadditive	Nanoadditive: size dosage	Base fuel	Base fuel properties					Nanofuel properties with respect to base fuel					References
			ρ_f	ν_f	T_{fp}	CN	H_f	ρ_f	ν_f	T_{fp}	CN	H_f	
			$\left(\frac{kg}{m^3}\right)$	$\left(\frac{mm^2}{s}\right)$	(°C)	(/)	$\left(\frac{MJ}{kg}\right)$	Maximum % increase (↑) or decrease (↓)					
Al_2O_3 NPs	20–30 nm 100 ppm	D100	841	4.56	60–80	56	48.6	↑ 38	↓ 9.6	–	↓ 1.1	↑ 2.5	[13]
	27–43 nm 100 ppm	D100	834	3.6	60	54	42–45	↑ 0.12	≈	↑ 15	↑ 1.5	–	[28, 32]
	≤50 nm 100 ppm	D100	825–855	10.6	42	48–57	42.5	–	↓ 8.3	↑ 69	–	↑ 0.5	[38, 48]
	<100 nm 0.006wt%	D100	834	3.6	60	48–57	42–45	↑ 0.02	↓ 5.4	↑ 11.8	–	–	[32, 53]
	45 nm 150 ppm	D100	844	2.3	60	48–57	42–45	↑ 2.6	–	↑ 10	↑ 5.2	–	[48, 56]
	30–50 nm 40 ppm	D/*ci*B20	843	3.56	69	54	40.9	↑ 1.7	↑ 2.2	↓ 7.2	↑ 1.3	↑ 1.3	[6]
	50 nm 100 ppm	D/*gs*B20	841	5.7	39	46	37	↑ 2.7	↑ 1.8	↑ 77	↑ 6.5	↑ 6.9	[41]
	<50 nm 0.5 g/L	D/*j*B20	800	–	57	–	41.7	↑ 0.3	–	↑ 1.8	–	↑ 0.3	[65]
	20–50 nm 30 mg/L	D/*jj*B20	847	4.06	–	52	45.4	≈	↑ 7.9	–	↑ 9.6	↑ 0.1	[24]
	32–48 nm 100 ppm	D/*m*B20	826	3.4	76	49	41.6	↑ 0.4	↓ 2.1	↓ 14.5	↑ 4.1	↑ 0.2	[1]

(continued)

Table 4.15 (continued)

Nanoadditive	Nanoadditive: size dosage	Base fuel	Base fuel properties					Nanofuel properties with respect to base fuel					References
			ρ_f	ν_f	T_{fp}	CN	H_f	ρ_f	ν_f	T_{fp}	CN	H_f	
			$\left(\frac{kg}{m^3}\right)$	$\left(\frac{mm^2}{s}\right)$	(°C)	(/)	$\left(\frac{MJ}{kg}\right)$	Maximum % increase (↑) or decrease (↓)					
	30–50 nm 100 ppm	D/*m*B25	853	3.2	65	–	38.5	↑ 0.1	↑ 17.2	↑ 1.5	–	↑ 1.9	[79]
	<100 nm 100 ppm	D/*pp*B25	843	4.4	68	50	41.5	↑ 0.6	↓ 2.0	↓ 14.7	↑ 4.0	↑ 0.1	[93]
	20 nm 60 ppm	D/*ts*B20	845	3.88	75	51	41.7	↑ 0.4	↑ 0.8	↑ 4.1	↑ 10	↑ 0.1	[74]
	50 nm 90 ppm	D/*wco*B10	835	5.85	–	51	42.7	↑ 0.6	↑ 4	–	↑ 6.1	↓ 0.2	[32]
	<100 nm 25 ppm	D/*zj*B25	846	3.56	56	55	–	↑ 0.4	↓ 5	↑ 1.8	↑ 3.6	–	[2]
	<100 nm 10 wt%	*cns*B100	844	4.28	–	51	37.5	↑ 1.4	↑ 0.9	–	↑ 0.04	↑ 0.9	[73]
	20 nm 50 mg/L	D/*j*B20/E10	840	2.86	20	53	40	↓ 0.4	↓ 10	↑ 10	↑ 1.9	↓ 2.1	[100]
	27–43 nm 2 wt%	GTL100	750	1.34	–	–	–	↑ 1.7	↑ 1.5	–	–	–	[39]

(continued)

Table 4.15 (continued)

Nanoadditive	Nanoadditive: size dosage	Base fuel	Base fuel properties					Nanofuel properties with respect to base fuel					References
			ρ_f	ν_f	T_{fp}	CN	H_f	ρ_f	ν_f	T_{fp}	CN	H_f	
			$\left(\frac{kg}{m^3}\right)$	$\left(\frac{mm^2}{s}\right)$	(°C)	(/)	$\left(\frac{MJ}{kg}\right)$	Maximum % increase (↑) or decrease (↓)					
AlOOH NPs	15 nm 200 ppm	D/*co*B20	851	5.24	62	49.6	36.8	↑ 30	↑ 24	↑ 19	↓ 3	↑ 3.5	[72]

Table 4.16 Properties of nanofuels with NMs based on cerium

Nanoadditive	Nanoadditive: size dosage	Base fuel	Base fuel properties					Nanofuel properties with respect to base fuel					References
			ρ_f	ν_f	T_{fp}	CN	H_f	ρ_f	ν_f	T_{fp}	CN	H_f	
			$\left(\frac{kg}{m^3}\right)$	$\left(\frac{mm^2}{s}\right)$	(°C)	(/)	$\left(\frac{MJ}{kg}\right)$	Maximum % increase (↑) or decrease (↓)					
CeO_2 NPs	15–30 nm 25 mg/L	D100	825	3.3	49	–	40.9	↑ 3.6	↑ 12.1	↑ 12.2	–	↓ 2	[26]
	50 nm 25 ppm	D100	848.6	5.80	62	47	45.1	↑ 0.1	↓ 10	↑ 19	↓ 6.4	↑ 0.4	[59]
	16 nm 30 ppm	*lg*O/W5	906	4.67	74	46	35.8	↑ 1.1	↑ 6.8	↓ 9.5	↑ 5.4	↑ 1.1	[7]
	<100 nm 50 ppm	D/*cs*B10	827	4.29	60	–	43.5	↑ 1.0	↑ 23.3	↑ 6.7	–	↑ 5.1	[66]
	<100 nm 40 mg/L	D/*cs*B20	–	–	87	–	41.1	–	–	↓ 6.9	–	↑ 0.3	[76]
	<100 nm 20 ppm	D/*lg*B25	848.5	3.38	50	48	42.1	↓ 1.2	↓ 8.4	↑ 5	↓ 0.5	↑ 1.7	[19]
	<25 nm 150 ppm	D/*m*B20	831	3.36	61	–	43.8	↓ 0.5	↑ 15.2	↑ 14.8	–	↓ 8.3	[46]
	<100 nm 90 ppm	D/*r*B5/W4	851.7	3.55	76	–	40.5	↑ 0.3	↑ 8.2	↑ 3.9	–	↓ 0.2	[37]
	<100 nm 90 ppm	D/*wco*B5/W3	–	3.6	78	–	44.3	–	↑ 6.1	↑ 2.6	↑	↓ 1.9	[44]
	15–30 nm 90 ppm	D/*wco*B5/W3	839	3.6	–	–	44.3	≈	↓ 1	–	–	↓ 1.1	[3, 4]

(continued)

Table 4.16 (continued)

Nanoadditive	Nanoadditive: size dosage	Base fuel	Base fuel properties					Nanofuel properties with respect to base fuel					References
			ρ_f	ν_f	T_{fp}	CN	H_f	ρ_f	ν_f	T_{fp}	CN	H_f	
			$\left(\frac{kg}{m^3}\right)$	$\left(\frac{mm^2}{s}\right)$	(°C)	(/)	$\left(\frac{MJ}{kg}\right)$	Maximum % increase (↑) or decrease (↓)					
	<100 nm 40 ppm	*j*B100	870–880	5.3–5.7	160–170	52–53	38–39	–	↑	↑	–	–	[65, 85]
	20–30 nm 100 ppm	*j*B/*tp*O10	868.7	6.39	–	–	41.7	≈	↑ 0.1	–	–	↓ 4.3	[92]
	15 nm 30 ppm	*no*B/W15	906	4.67	74	–	35.8	↑ 1.1	↑ 6.9	↓ 9.5	–	↑ 1.1	[19]

Table 4.17 Properties of nanofuels with NMs based on cobalt

Nanoadditive	Nanoadditive: size dosage	Base fuel	Base fuel properties					Nanofuel properties with respect to base fuel					References
			ρ_f	ν_f	T_{fp}	CN	H_f	ρ_f	ν_f	T_{fp}	CN	H_f	
			$\left(\frac{kg}{m^3}\right)$	$\left(\frac{mm^2}{s}\right)$	(°C)	(/)	$\left(\frac{MJ}{kg}\right)$	Maximum % increase (↑) or decrease (↓)					
Co_3O_4 NPs	<100 nm 100 ppm	D/*gs*B20	841	5.72	39	46	37	↑ 1.5	↑ 2.6	↑ 74	↑ 6.5	↑ 6.0	[42]
	30 nm 50 ppm	D/*wfo*B20	830	3.1	140	50	41.6	–	–	–	↑ 1.2	↑	[49]

Table 4.18 Properties of nanofuels with NMs based on copper

Nanoadditive	Nanoadditive: size dosage	Base fuel	Base fuel properties					Nanofuel properties with respect to base fuel					References
			ρ_f	ν_f	T_{fp}	CN	H_f	ρ_f	ν_f	T_{fp}	CN	H_f	
			$\left(\frac{kg}{m^3}\right)$	$\left(\frac{mm^2}{s}\right)$	(°C)	(/)	$\left(\frac{MJ}{kg}\right)$	Maximum % increase (↑) or decrease (↓)					
CuO NPs	30–50 nm 50 ppm	D100	833.5	3.6	60	54	42–45	↑ 0.1	↓ 2.8	↑ 10	↑ 1.1	–	[28, 48]
	<100 nm 0.006 wt%	D100	834.1	3.6	59.5	48–57	42–45	↑	↓ 3.0	↑ 13	–	–	[48, 53]
	<100 nm 0.5 wt%	D100	833	3.6	60	48–57	42	↑ 0.1	↓ 2.8	↑ 10	–	↑ 1.0	[48, 84]
	32 nm 50 ppm	D/W10	852	9.2	–	–	–	↑ 0.2	≈	–	↑ 1.2	–	[29]
	<50 nm 50 mg/L	D/*pp*B20	723	3.45	63	–	36.3	↑ 2.4	↑ 8.1	↑ 8.5	–	↑ 6	[16]
	50 nm 50 ppm	D/*pp*B20	824	3.02	69	–	43.7	↑ 2.6	↑ 60	↓ 4.3	–	↑ 0.3	[67]

Table 4.19 Properties of nanofuels with NMs based on iron

Nanoadditive	Nanoadditive:size dosage	Base fuel	Base fuel properties					Nanofuel properties with respect to base fuel					References
			ρ_f	ν_f	T_{fp}	CN	H_f	ρ_f	ν_f	T_{fp}	CN	H_f	
			$\left(\frac{kg}{m^3}\right)$	$\left(\frac{mm^2}{s}\right)$	(°C)	(/)	$\left(\frac{MJ}{kg}\right)$	Maximum % increase (↑) or decrease (↓)					
γ-Fe_2O_3 NPs	35 nm 100 ppm	D100	825–855	10.6	42	48–57	42.5	–	↓ 6.8	↑ 64	–	↑ 0.5	[38, 48]
	30 nm 1 wt%	D/*pp*B20	859	3.4	81	51	41.6	↑ 2.4	↑ 6.8	≈	↑ 2.1	↑ 0.2	[45]
	<100 nm 100 ppm	D/*pp*B20	843	3.5	76	49	41.6	↑ 0.6	↑ 17	↓ 31	↑ 14	↑ 2.5	[52]
Fe_3O_4 NPs	10 nm 1 wt%	*rb*B100	882	6.32	138	52	38.1	↑ 3.2	↑ 3.6	↑ 4.3	↑ 5.8	↑ 2.7	[75]

Table 4.20 Properties of nanofuels with NMs based on magnesium

Nanoadditive	Nanoadditive: size dosage	Base fuel	Base fuel properties					Nanofuel properties with respect to base fuel					References
			ρ_f	ν_f	T_{fp}	CN	H_f	ρ_f	ν_f	T_{fp}	CN	H_f	
			$\left(\frac{kg}{m^3}\right)$	$\left(\frac{mm^2}{s}\right)$	(°C)	(/)	$\left(\frac{MJ}{kg}\right)$	Maximum % increase (↑) or decrease (↓)					
MgO NPs	33 nm 50 ppm	D/W10	852	9.2	–	–	–	↑ 0.2	≈	–	↑ 1.2	–	[29]
	20–30 nm 30 ppm	D/*wco*B20	826	2.67	98	–	29.6	↑ 0.4	↑ 0.7	↑ 5.1	–	↑ 0.9	[80]
	10–30 nm 50 ppm	*r*B100	896	5.09	> 180	51–55	40–42	↑ 0.1	↓ 5.1	≈	–	–	[43, 61]
	20–30 nm 30 ppm	*wco*B100	874	3.23	150	–	25.8	↑ 1.0	↑ 0.7	↑ 2.7	–	↑ 0.7	[80]

Table 4.21 Properties of nanofuels with NMs based on titanium

Nanoadditive	Nanoadditive: size dosage	Base fuel	Base fuel properties					Nanofuel properties with respect to base fuel					References
			(/)	ν_f	T_{fp}	CN	H_f	ρ_f	ν_f	T_{fp}	CN	H_f	
			$\left(\frac{kg}{m^3}\right)$	$\left(\frac{mm^2}{s}\right)$	(°C)	(/)	$\left(\frac{MJ}{kg}\right)$	Maximum % increase (↑) or decrease (↓)					
TiO_2 NPs	45 nm 150 ppm	D100	844	2.3	60	48–57	42–45	↑ 2.7	–	↑ 11	↑ 3.3	–	[48, 56]
	10–20 nm 80 mg/L	D100	697.2	2.98	51	–	41.8	↑ 0.4	↑ 6.2	↑ 41	–	↑ 0.6	[14]
	10–50 nm 40 ppm	D/W5	855	5.35	64	42	40.3	↑ 1.2	↑ 1.3	↑ 3.1	↓ 0.7	↑ 3.3	[96]
	10–20 nm 25 ppm	D/*me*B20	843	2.76	76	53	43.6	↑ 0.3	↑ 13.4	↑ 7.8	↓ 1.9	↑ 0.2	[47]
	10–50 nm 40 ppm	*p*O/W5	830	5.15	60	43	40.3	↑ 1.1	↑ 2.1	↑ 8.3	↑ 1.9	↑ 2.4	[96]
	<100 nm 40 ppm	D/*ci*B20	843.3	3.56	68	54	41.7	↑ 0.1	↑ 4.5	↓ 5.9	↑ 0.2	↑ 0.6	[71]
	<100 nm 0.01 wt%	D/*wco*B20	842.1	2.79	71	–	40.8	↑ 0.02	↑ 4.1	↑ 2.5	–	↑ 0.6	[60]
	20 nm 100 ppm	*ci*B100	868.6	4.72	122	52	38	↑ 0.2	↑ 0.6	↑ 1.6	↑ 5.8	↓ 1.2	[10]
	20 nm 100 ppm	*ci*B100	868.6	4.72	122	52	38	↑ 0.2	↑ 0.6	↑ 1.6	↑ 5.8	↓ 1.2	[55]
	60 nm 200 ppm	*m*B100	844	4.44	–	50	39.1	↑ 7.5	↑ 6.3	–	↑ 6	↓ 2.3	[62]
	50 nm 200 ppm	*mo*B100	864	4.3	–	52	38.1	↑ 3.1	↑ 1.9	–	↑ 9.6	↓ 1.2	[104]

(continued)

Table 4.21 (continued)

Nanoadditive	Nanoadditive: size dosage	Base fuel	Base fuel properties					Nanofuel properties with respect to base fuel					References
			(/)	ν_f	T_{fp}	CN	H_f	ρ_f	ν_f	T_{fp}	CN	H_f	
			$\left(\frac{kg}{m^3}\right)$	$\left(\frac{mm^2}{s}\right)$	(°C)	(/)	$\left(\frac{MJ}{kg}\right)$	Maximum % increase (↑) or decrease (↓)					
	<100 nm 0.01 wt%	D/*wco*B20/Bu10	840.1	2.62	47	–	40	↑ 0.01	↑ 0.5	↓ 3.7	–	↓ 0.3	[60]

Table 4.22 Properties of nanofuels with NMs based on zinc

Nanoadditive	Nanoadditive: size dosage	Base fuel	Base fuel properties					Nanofuel properties with respect to base fuel					References
			ρ_f	ν_f	T_{fp}	CN	H_f	ρ_f	ν_f	T_{fp}	CN	H_f	
			$\left(\frac{kg}{m^3}\right)$	$\left(\frac{mm^2}{s}\right)$	(°C)	(/)	$\left(\frac{MJ}{kg}\right)$	Maximum % increase (↑) or decrease (↓)					
ZnO NPs	67 nm 100 ppm	D100	825–855	10.6	42	48–57	42.5	–	↓ 7.6	↑ 66.7	–	↑ 3.1	[38, 48]
	34 nm 50 ppm	D/W10	852	9.2	–	–	–	↑ 0.2	↓ 1.1	–	↑ 1.2	–	[29]
	35 nm 100 ppm	*ci*B100	868.6	4.72	122	52	38	↑ 0.4	↑ 1.3	↑ 3.3	↑ 7.7	↓ 1.8	[55]
	50 nm 100 ppm	*ci*B100	868.6	4.72	122	52	38	≈	↑ 0.8	↓ 1.7	↑ 5.8	↓ 1.2	[9]
	30 nm 250 ppm	D/*p*B30/E30	828.5	2.42	10	57	34.9	↑ 0.9	↓ 4.1	↑ 60	↑	↑ 5.6	[68]

Table 4.23 Properties of nanofuels with NMs with C NTs

Nanoadditive	Nanoadditive:size dosage	Base fuel	Base fuel properties					Nanofuel properties with respect to base fuel					References
			ρ_f	ν_f	T_{fp}	CN	H_f	ρ_f	ν_f	T_{fp}	CN	H_f	
			$\left(\frac{kg}{m^3}\right)$	$\left(\frac{mm^2}{s}\right)$	(°C)	(/)	$\left(\frac{MJ}{kg}\right)$	Maximum % increase (↑) or decrease (↓)					
C NTs	40–60 nm 100 ppm	D100	841	4.56	60–80	56	48.6	↑ 1.1	↓ 16	–	↓ 2	↑ 5.5	[13]
	<100 nm 100 ppm	D100	830.8	2.41	62	52	42.9	↑ 0.4	↑ 1.2	↑ 3.2	↑ 1.1	↑ 0.5	[98]
	<100 nm 50 ppm	D/*h*B20	–	5.7	170	–	36	–	↓ 5.3	↓ 3.5	–	↓ 2.5	[86]
	<100 nm 100 ppm	D/*s*B50	851.4	3.14	98	55	40.7	↑ 0.4	↑ 0.6	↑ 2.1	↑ 1.3	↑ 0.2	[98]
	33 nm 60 ppm	D/*ts*B20	843	3.86	74	48	41.7	↑ 0.3	↑ 1.0	↑ 5.4	↑ 8.3	↑ 0.1	[74]
	4–8 nm 60 ppm	D/*wco*B2	820.7	2.31	–	–	42.7	↑ 0.5	↑ 14.2	–	–	↓ 2.3	[30]
	<100 nm 100 ppm	*s*B100	871.2	3.92	129	56	37.9	↑ 0.4	↑ 0.8	↑ 2.3	↑ 1.1	↑ 0.5	[98]
	<200 nm 100 ppm	*j*B/*tp*O10	868.7	6.386	–	–	41.7	↑ 0.4	↓ 17.8	–	–	↓ 6.5	[92]
	16 nm 100 ppm	*j*B/W5	899.8	5.4	140	51	37.1	≈	↑ 9.4	↓ 12.9	↑ 9.8	↑ 2.2	[11]
	5–20 nm 100 ppm	*wco*B100	870–890	4.6–4.8	170–180	57–63	37–39	–	–	–	↑ 4.8	–	[3, 54]

Table 4.24 Properties of nanofuels with NMs based on carbon, graphite, and graphene

Nanoadditive	Nanoadditive: size dosage	Base fuel	Base fuel properties					Nanofuel properties with respect to base fuel					References
			ρ_f	ν_f	T_{fp}	CN	H_f	ρ_f	ν_f	T_{fp}	CN	H_f	
			$\left(\frac{kg}{m^3}\right)$	$\left(\frac{mm^2}{s}\right)$	(°C)	(/)	$\left(\frac{MJ}{kg}\right)$	Maximum % increase (↑) or decrease (↓)					
swC NTs	1–2 nm 25 ppm	D100	848.6	5.80	62	47	45.1	↑ 0.1	↓ 8.4	↑ 170	↑ 12.8	↑ 1.0	[59]
mwC NTs	10–20 nm 90 ppm	D/*wco*B10	834	5.85	–	57	42.7	↓ 0.3	↓ 0.3	–	↑ 0.5	↑ 0.1	[33]
	10–15 nm 50 mg/L	*j*B/Bu40	849.9	3.73	–	44	37.5	↑ 0.2	↓ 1.1	–	↑ 10.2	↑ 0.1	[20]
G'O NPs	1–2 nm 25 ppm	D100	848.8	4.74	83	46	42–45	↑ 0.05	↓ 9.1	↑ 140	↑ 8.5	–	[33, 59]
G NPTs	6–8 nm 100 mg/L	D/*j*B20	850.1	4.05	–	52	41.1	≈	↑ 26.7	–	↑ 11.2	↑ 0.2	[22]
	6–8 nm 50 mg/L	*j*B/Bu40	849.9	3.73	–	44	37.5	↑ 0.1	↓ 1.3	–	↑ 10.2	↑ 0.1	[20]
GO NPs	6 nm 50 mg/L	*j*B100	883	5.26	–	52	40.6	↑ 0.1	↓ 2.5	–	↑ 4.8	↑ 0.2	[23]
	1–2 nm 90 ppm	D/*aa*B20	838.1	5.56	–	47	41.9	≈	↓ 0.6	–	↑ 3.2	↑ 0.5	[31]

Table 4.25 Properties of nanofuels with other NMs

Nanoadditive	Nanoadditive: size dosage	Base fuel	Base fuel properties					Nanofuel properties with respect to base fuel					References
			ρ_f	ν_f	T_{fp}	CN	H_f	ρ_f	ν_f	T_{fp}	CN	H_f	
			H_f	$\left(\frac{mm^2}{s}\right)$	(°C)	(/)	$\left(\frac{MJ}{kg}\right)$	Maximum % increase (↑) or decrease (↓)					
B_2O_3 NPs	80 nm 0.5 wt%	D100	825–855	1.3–4.1	60–80	48–57	42–45	–	–	–	–	↑	[50]
MnO NPs	33 nm 50 ppm	D/W10	852	9.2	–	–	–	↑ 0.2	≈	–	↑ 1.2	–	[29]
Mn_2O_3 NPs	30 nm 50 ppm	D/*wfo*B20	830	3.1	140	50	41.6	–	–	–	↑ 1.2	↑	[49]
SiO_2 NPs	20–30 nm 100 ppm	D100	841	4.56	60–80	56	48.6	↓ 0.6	↓ 6.6	–	↓ 0.5	↑ 0.04	[13]
	10–30 nm 50 ppm	*r*B100	896	5.09	> 180	51–55	40–42	≈	↓ 4.9	≈	–	–	[43, 61]
ZnO NPs	67 nm 100 ppm	D100	825–855	10.6	42	48–57	42.5	–	↓ 7.6	↑ 66.7	–	↑ 3.1	[38, 48]
	34 nm 50 ppm	D/W10	852	9.2	–	–	–	↑ 0.2	↓ 1.1	–	↑ 1.2	–	[29]
	35 nm 100 ppm	*ci*B100	868.6	4.72	122	52	38	↑ 0.4	↑ 1.3	↑ 3.3	↑ 7.7	↓ 1.8	[55]
	50 nm 100 ppm	*ci*B100	868.6	4.72	122	52	38	≈	↑ 0.8	↓ 1.7	↑ 5.8	↓ 1.2	[9]
	30 nm 250 ppm	D/*p*B30/E30	828.5	2.42	10	57	34.9	↑ 0.9	↓ 4.1	↑ 60	↑	↑ 5.6	[68]

(continued)

Table 4.25 (continued)

Nanoadditive	Nanoadditive: size dosage	Base fuel	Base fuel properties					Nanofuel properties with respect to base fuel					References
			ρ_f	ν_f	T_{fp}	CN	H_f	ρ_f	ν_f	T_{fp}	CN	H_f	
			H_f	$\left(\frac{mm^2}{s}\right)$	(°C)	(/)	$\left(\frac{MJ}{kg}\right)$	Maximum % increase (↑) or decrease (↓)					
Al_2O_3 and CeO_2 NPs	51 and 32 nm 30 ppm	D/*j*B20	843	2.58	55	–	41.7	↑ 0.1	↑ 0.4	↓ 5.5	–	↑ 1.2	[69, 70]
	51 and 32 nm 30 ppm	*j*B100	873	4.1	85	–	39.5	↑ 1.1	≈	↓ 2.4	–	↑ 1.8	[69, 70]
SiO_2 and TiO_2 NPs	50 nm 100 ppm	D/*bba*B20	816	3.12	60	47	40.4	↑ 0.1	↓ 3.5	↑ 5.2	↑ 2.1	↑ 8.9	[40]

4.2.11 Nanofuels with NMs Based on Carbon, Graphite, and Graphene

Addition of carbon, graphite, and graphene-based NMs to all tested base fuels increases the flash point, cetane number, and calorific value, Table 4.24. The effects on fuel density and viscosity vary in dependence on the NMs size and dosage [20, 22, 23, 31, 33, 59].

4.2.12 Nanofuels with Other NMs

In Table 4.25, the effects of addition of B_2O_3, MnO, Mn_2O_3, SiO_2, ZnO, Al_2O_3 and CeO_2 NPs, and SiO_2 and TiO_2 NPs to various base fuels are presented. The addition of B_2O_3 NPs to D100 increases the calorific value of the fuel [50]. The reported influences of Mn-based NMs on nanofuels properties are related to density, cetane number, and calorific values and these values are usually increased [29, 49]. By added SiO_2 NPs, practically all measured nanofuel properties are decreased [13, 43, 61]. The addition of a mixture of two NPs, Al_2O_3 and CeO_2 NPs [69, 70] and SiO_2 and TiO_2 NPs [40] increases the density, cetane number, and calorific value, meanwhile their effects on viscosity and flash point vary with respect to the base fuel and NPs size and dosage.

As can be seen from Tables 4.14, 4.15, 4.16, 4.17, 4.18, 4.19, 4.20, 4.21, 4.22, 4.23, 4.24, and 4.25, the NMs addition to a base fuel influences notably the fuel properties. These effects depend on the base fuel, NMs type, NMs size, NMs dosage, engaged preparation method, and used surfactants. Based on the data given in Tables 4.14, 4.15, 4.16, 4.17, 4.18, 4.19, 4.20, 4.21, 4.22, 4.23, 4.24, and 4.25, one can say that, in general, the addition of NMs usually increases ρ_f, ν_f, CN, and H_f, although these effects may vary notably for a particular property. Furthermore, NMs additives can improve thermal conductivity and evaporation rate of a fuel [94]. All these property variations, induced by addition of NMs, have a very important role in the combustion process in a diesel engine. Apart from this, Al_2O_3 and Co_3O_4 NPs in D/*gs*B20 fuel as well as γ-Fe_2O_3, ZnO, swC, and G'O NMs in D100 significantly increase T_{fp}, making these nanofuels safer in storage and transit.

Finally, an analysis of Tables 4.14, 4.15, 4.16, 4.17, 4.18, 4.19, 4.20, 4.21, 4.22, 4.23, 4.24, and 4.25, also shows that by using adequate preparation methods it is possible to get surfactant-free nanofuels that exhibit practically the same properties as the nanofuels prepared with a surfactant. For example, by adding 100 ppm of 20 nm-sized TiO_2 NPs to *ci*B100, the fuel properties [55] are practically the same as those of the similar nanofuel prepared by using the Span 80^{TM} surfactant [9].

References

1. Aalam CS, Saravanan CG (2017) Effects of nano metal oxide blended Mahua biodiesel on CRDI diesel engine. Ain Shams Eng J 8:689–696
2. Aalam CS, Saravanan CG, Kannan M (2015) Experimental investigations on a CRDI system assisted diesel engine fueled with aluminum oxide nanoparticles blended biodiesel. Alex Eng J 54:351–358
3. Aghbashlo M, Tabatabaei M, Khalife E, Najafi B, Mirsalim SM, Gharehghani A, Mohammadi P, Dadak A, Shojaei TR, Khounani Z (2017) A novel emulsion fuel containing aqueous nano cerium oxide additive in diesel-biodiesel blends to improve diesel engines performance and reduce exhaust emissions: part II—exergetic analysis. Fuel 205:262–271
4. Aghbashlo M, Tabatabaei M, Khalife E, Shojaei TR, Dadak A (2018) Exergoeconomic analysis of a DI diesel engine fueled with diesel/biodiesel (B5) emulsions containing aqueous nano cerium oxide. Energy 149:967–978
5. Ali HM, Babar H, Shah TR, Sajid MU, Qasim MA, Javed S (2018) Preparation techniques of TiO_2 nanofluids and challenges: a review. Appl Sci 8:587
6. Anchupogu P, Rao LN, Banavathu B (2018) Effect of alumina nano additives into biodiesel-diesel blends on the combustion performance and emission characteristics of a diesel engine with exhaust gas recirculation. Environ Sci Pollut Res 25:23294–23306
7. Annamalai M, Dhinesh B, Nanthagopal K, Sivaramakrishnan P, Lalvani JIJ, Parthsarathy M, Annamalai K (2016) An assessment on performance, combustion and emission behavior of a diesel engine powered by ceria nanoparticle blended emulsified biofuel. Energy Convers Manag 123:372–380
8. Ashok B, Nanthagopal K, Mohan A, Johny A, Tamilarasu A (2017) Comparative analysis on the effect of zinc oxide and ethanox as additives with biodiesel in CI engine. Energy 140:352–364
9. Ashok B, Nanthagopal K, Subbarao R, Johny A, Mohan A, Tamilarasu A (2017) Experimental studies on the effect of metal oxide and antioxidant additives with *Calophyllum inophyllum* methyl ester in compression ignition engine. J Clean Prod 166:474–484
10. Ashok B, Nanthagopal K, Vignesh DS (2018) *Calophyllum inophyllum* methyl ester biodiesel blend as an alternate fuel for diesel engine applications. Alex Eng J 57(3):1239–1247
11. Basha JS, Anand RB (2014) Performance, emission and combustion characteristics of a diesel engine using carbon nanotubes blended Jatropha methyl ester emulsions. Alex Eng J 53:259–273
12. Chaichan MT, Kadhum AAH, Al-Amiery AA (2017) Novel technique for enhancement of diesel fuel: Impact of aqueous alumina nano-fluid on engine's performance and emissions. Case Stud Therm Eng 10:611–620
13. Chen AF, Adzmi MA, Adam A, Othman MF, Kamaruzzaman MK, Mrwan AG (2018) Combustion characteristics, engine performances and emissions of a diesel engine using nanoparticle-diesel fuel blends with aluminium oxide, carbon nanotubes and silicon oxide. Energy Convers Manag 171:461–477
14. D'Silva R, Binu KG, Thirumaleshwara B (2015) Performance and Emission characteristics of a CI engine fueled with diesel and TiO_2 nanoparticles as fuel additive. Mater Today Proc 2:3728–3735
15. D'Silva R, Vinoothan K, Binu KG, Thirumaleshwara B, Raju K (2016) Effect of titanium dioxide and calcium carbonate nanoadditives on the performance and emission characteristics of CI engine. J Mech Eng Autom 6(5A):28–31
16. D'Silva R, Hafeez M, Fernandez J, Paloth F, Rahiz IA, Binu KG, Raju K, Thirumaleshwara B (2017) Effect of copperoxide nanoadditives on the performance and emissions characteristics of a CI engine. Energy and Power 7(4):99–104
17. Devarajan Y, Munuswamy DB, Mahalingam A (2018) Influence of nano-additive on performance and emission characteristics of a diesel engine running on neat neem oil biodiesel. Environ Sci Pollut Res 25:26167–26172

18. Devarajan Y, Mahalingam A, Munuswamy DB, Arunkumar T (2018) Combustion, performance, and emission study of a research diesel engine fueled with palm oil biodiesel and its additive. Energy Fuels 32:8447–8452
19. Dhinesh B, Annamalai M (2018) A study on performance, combustion and emission behaviour of diesel engine powered by novel nano nerium oleander biofuel. J Clean Prod 196:74–83
20. El-Seesy AI, Hassan H (2019) Investigation of the effect of adding graphene oxide, graphene nanoplatelet, and multiwalled carbon nanotube additives with n-butanol-Jatropha methyl ester on a diesel engine performance. Renew Energy 132:558–574
21. El-Seesy AI, Abdel-Rahman AK, Bady M, Ookawara S (2016) The influence of multi-walled carbon nanotubes additives into non-edible biodiesel-diesel fuel blend on diesel engine performance and emissions. Energy Proc 100:166–172
22. El-Seesy A, Hassan H, Ookawara S (2018) Effects of graphene nanoplatelet addition to jatropha Biodiesel-Diesel mixture on the performance and emission characteristics of a diesel engine. Energy 147:1129–1152
23. El-Seesy AI, Hassan H, Ookawara S (2018) Performance, combustion, and emission characteristics of a diesel engine fueled with Jatropha methyl ester and graphene oxide additives. Energy Convers Manag 166:674–686
24. El-Seesy AI, Attia AMA, El-Battsh M (2018) The effect of aluminum oxide nanoparticles addition with Jojoba methyl ester-diesel fuel blend on a diesel engine performance, combustion and emission characteristics. Fuel 224:147–166
25. Ettefaghi E, Ghobadian B, Rashidi A, Najafi G, Khoshtaghaza MH, Rashtchi M, Sadeghian S (2018) A novel bio-nano emulsion fuel based on biodegradable nanoparticles to improve diesel engines performance and reduce exhaust emissions. Renew Energy 125:64–72
26. Feroskhan M, Ismail S, Gosavi S, Tankhiwale P, Khan Y (2019) Optimization of performance and emissions in a biogas-diesel dual fuel engine with cerium oxide nanoparticle addition. Proc Inst Mech Eng Part D J Automob Eng 233(5):1178–1193
27. Ghafoori M, Ghobadian B, Najafi G, Layeghi M, Rashidi A, Mamat R (2015) Effect of nano-particles on the performance and emission of a diesel engine using biodiesel-diesel blend. Int J Autom Mech Eng 12:3097–3108
28. Gumus S, Ozcan H, Ozbey M, Topaloglu B (2016) Aluminum oxide and copper oxide nanodiesel fuel properties and usage in a compression ignition engine. Fuel 163:80–87
29. Hasannuddin AK, Yahya WJ, Sarah S, Ithmin AM, Syahrullail S, Sidik NAC, Kassim KAA, Ahmad Y, Hirofumi N, Ahmad MA, Sugeng DA, Zuber MA, Ramlan NA (2018) Nano-additives incorporated water in diesel emulsion fuel: fuel properties, performance and emission characteristics assessment. Energy Convers Manag 169:291–314
30. Heydari-Maleney K, Taghizadeh-Alisaraei A, Ghobadian B, Abbaszadeh-Mayvan A (2017) Analyzing and evaluation of carbon nanotubes additives to diesohol-B2 fuels on performance and emission of diesel engines. Fuel 196:110–123
31. Hoseini SS, Najafi G, Ghobadian B, Mamat R, Ebadi MT, Yusaf T (2018) Novel environmentally friendly fuel: The effects of nanographene oxide additives on the performance and emission characteristics of diesel engines fuelled with *Ailanthus altissima* biodiesel. Renew Energy 125:283–294
32. Hosseini SH, Taghizadeh-Alisaraei A, Ghobadian B, Abbaszadeh-Mayvan A (2017) Effect of added alumina as nano-catalyst to diesel-biodiesel blends on performance and emission characteristics of CI engine. Energy 124:543–552
33. Hosseini SH, Taghizadeh-Alisaraei A, Ghobadian B, Abbaszadeh-Mayvan A (2017) Performance and emission characteristics of a CI engine fuelled with carbon nanotubes and diesel-biodiesel blends. Renew Energy 111:201–213
34. Jayanthi P, Srinivasa RM (2016) Effect of nanoparticles additives on performance and emissions characteristics of a DI diesel engine fueled with biodiesel. Int J Adv Eng Technol 9(6):689–695
35. Jeryrajkumar L, Anbarasu G, Elangovan T (2016) Effects on nano additives on performance and emission characteristics of *Calophyllim inophyllum* biodiesel. Int J Chem Tech Res 9(4):210–219

36. Jiang L, Wang YD, Roskilly AP, Xie XL, Zhang ZC, Wang RZ (2018) Investigation on thermal properties of a novel fuel blend and its diesel engine performance. Energy Convers Manag 171:1540–1548
37. Jiaqiang E, Zhang Z, Chen J, Pham M, Zhao X, Peng Q, Zhang B, Yin Z (2018) Performance and emission evaluation of a marine diesel engine fueled by water biodiesel-diesel emulsion blends with a fuel additive of a cerium oxide nanoparticle. Energy Convers Manag 169:194–205
38. Kamesh MR, Madhu D (2017) Effect of nanoparticles on the emissions of a CI engine. Int J Renew Energy Res 7(2):751–757
39. Kannaiyan K, Sadr R (2017) The effects of alumina nanoparticles as fuel additives on the spray characteristics of gas-to-liquid jet fuels. Exp Thermal Fluid Sci 87:93–103
40. Karthikeyan S, Prathima A (2017) Environmental effect of CI engine using microalgae methyl ester with doped nano additives. Transp Res Part D 50:385–396
41. Karthikeyan S, Elango A, Silaimani SM, Prathima A (2014) Role of Al_2O_3 nano additive in GSO biodiesel on the working characteristics of a CI engine. Indian J Chem Technol 21:285–289
42. Karthikeyan S, Elango A, Prathima A, Raja K (2014) Environmental effects of nano additive Co_3O_4 in Grape seed oil biofuel fueled in CI engine. Res J Chem Environ 18(5):14–18
43. Kegl B, Kegl M (2017) Experimental investigation review of biodiesel usage in bus diesel engine. Therm Sci 21(1):639–654
44. Khalife E, Tabatabei M, Najafi B, Mirsalim SM, Gharehghani A, Mohammadi P, Aghbashlo M, Ghaffari A, Khounani Z, Shojaei TR, Salleh MAM (2017) A novel emulsion fuel containing aqueous nano cerium oxide additive in diesel-biodiesel blends to improve diesel engines performance and reduce exhaust emissions: part I—experimental analysis. Fuel 207:741–750
45. Kumar S, Dinesha P, Bran I (2017) Influence of nanoparticles on the performance and emission characteristics of a biodiesel fuelled engine: an experimental analysis. Energy 140:98–105
46. Kumar MV, Babu AV, Kumar PR (2019) Influence of metal-based cerium oxide nanoparticle additive on performance, combustion, and emissions with biodiesel in diesel engine. Environ Sci Pollut Res 26:7651–7664
47. Lawrence KR, Tharigonda H, Alluru G (2018) An experimental investigation on combined effect of EGR and SCR for diesel engine fueled with the addition of TiO_2 nanoparticle to *M. Elangi* methyl ester blend. Aust J Mech Eng. https://doi.org/10.1080/14484846.2018.1519053
48. Mat SC, Idroas MY, Hamid MF, Zainal ZA (2018) Performance and emissions of straight vegetable oils and its blends as a fuel in diesel engine: a review. Renew Sustain Energy Rev 82:808–823
49. Mehregan M, Moghiman M (2018) Effects of nano-additives on pollutants emission and engine performance in urea-SCR equipped engine fueled with blended-biodiesel. Fuel 222:402–406
50. Mehta RN, Chakraborty M, Parikh PA (2014) Nanofuels: combustion, engine performance and emissions. Fuel 120:91–97
51. Mirzajanzadeh M, Tabatabaei M, Ardjmand M, Rashidi A, Ghobadian B, Barkhi M, Pazouki M (2015) A novel soluble nano-catalysts in diesel-biodiesel fuel blends to improve diesel engines performance and reduce exhaust emissions. Fuel 139:374–382
52. Muthusamy S, Nallathambi SS, Ramasamy RK, Mohamed ST (2018) Effects of nanoparticles blended biodiesel on single cylinder CI engine. Mater Today Proc 5:6831–6838
53. Naik V, Kumar KK (2018) Performance and emission characteristics of diesel engines with Al_2O_3 and CuO nanoparticles as additives. Int J Mech Eng Technol 9(2):791–798
54. Najafi G (2018) Diesel engine combustion characteristics using nano-particles in biodiesel-diesel blends. Fuel 212:668–678
55. Nanthagopal K, Ashok B, Tamilarasu A, Johny A, Mohan A (2017) Influence on the effect of zinc oxide and titanium dioxide nanoparticles as an additive with *Calophyllum inophyllum* methyl ester in a CI engine. Energy Convers Manag 146:8–19

56. Nassir AK, Shahad HAK (2018) Experimental study of effect of nanoparticles addition on combustion phasing in diesel engine. Int J Mech Mech Eng 18(1):87–97
57. Nayak S, Shet VB, Rao CV, Joshi K (2018) Performance evaluation and emission characteristics of a 4 stroke diesel engine using green synthesized silver nanoparticles blended biodiesel. Mater Today Proc 5:7889–7897
58. Nour M, El-Seesy AI, Abdel-Rahman AK, Bady M (2018) Influence of adding aluminum oxide nanoparticles to diesterol blends on the combustion and exhaust emission characteristics of a diesel engine. Exp Thermal Fluid Sci 98:634–644
59. Ooi JB, Ismail HM, Tan BT, Wang X (2018) Effects of graphite oxide and single-walled carbon nanotubes as diesel additives on the performance, combustion, and emission characteristics of a light-duty diesel engine. Energy 161:70–80
60. Örs I, Sarıkoç S, Atabani AE, Ünalan S, Akansu SO (2018) The effects on performance, combustion and emission characteristics of DICI engine fuelled with TiO_2 nanoparticles addition in diesel/biodiesel/n-butanol blends. Fuel 234:177–188
61. Özgür T, Özcanli M, Aydin K (2015) Investigation of nanoparticle additives to biodiesel for improvement of the performance and exhaust emissions in a compression ignition engine. Int J Green Energy 12:51–56
62. Pandian AK, Ramakrishnan RBB, Devarajan Y (2017) Emission analysis on the effect of nanoparticles on neat biodiesel in unmodified diesel engine. Environ Sci Pollut Res 24:23273–23278
63. Panithasan MS, Gopalakichenin D, Venkadesa G, Veeraraagavan S (2019) Impact of rice husk nanoparticle on the performance and emission aspects of a diesel engine running on blends of pine oil-diesel. Environ Sci Pollut Res 26(1):282–291
64. Paramashivaih BM, Banapurmath NR, Rajashekhar CR, Khandal SV (2018) Studies on effect of graphene nanoparticles addition in different levels with *Simarouba* biodiesel and diesel blends on performance, combustion and emission characteristics of CI engine. Arab J Sci Eng 43:4793–4801
65. Patel HK, Kumar S (2017) Experimental analysis on performance of diesel engine using mixture of diesel and bio-diesel as a working fuel with aluminum oxide nanoparticle additive. Therm Sci Eng Progr 4:252–258
66. Patil DS, Chopade DA, Kumbhalkar MA (2018) Experimental investigation of effect of cerium oxide nanoparticles as a fuel additive in cottonseed biodiesel blends. MAYFEB J Mech Eng 1:1–12
67. Perumal V, Ilangkumaran M (2018) The influence of copper oxide nano particle added pongamia methyl ester biodiesel on the performance, combustion and emission of a diesel engine. Fuel 232:791–802
68. Prabakaran B, Udhoji A (2016) Experimental investigation into effects of addition of zinc oxide on performance, combustion and emission characteristics of diesel-biodiesel-ethanol blends in CI engine. Alex Eng J 55:3355–3362
69. Prabu A (2017) Nanoparticles as additive in biodiesel on the working characteristics of a DI diesel engine. Ain Shams Eng J 9(4):2343–2349
70. Prabu A (2018) Engine characteristic studies by application of antioxidants and nanoparticles as additives in biodiesel diesel blends. J Energy Res Technol 140:1–7
71. Praveen A, Rao GLN, Balakrishna B (2018) Performance and emission characteristics of a diesel engine using *Calophyllum inophyllum* biodiesel blends with TiO_2 nanoadditives and EGR. Egypt J Petrol 27:731–738
72. Praveena V, Narayanan SA, Hariskousik G, Balavinayagam K (2018) Effect of aluminium hydroxide nano fluid in castor oil biodiesel fueled diesel engine. Mater Sci Eng 402:012036
73. Radhakrishnan S, Munuswamy DB, Devarajan Y, Arunkumar T, Mahalingam A (2018) Effect of nanoparticle on emission and performance characteristics of a diesel engine fueled with cashew nut shell biodiesel. Energy Sources A Recov Util Environ Eff 40(20):2485–2493
74. Raju VD, Kishore PS, Nanthagopal K, Ashok B (2018) An experimental study on the effect of nanoparticles with novel tamarind seed methyl ester for diesel engine applications. Energy Convers Manag 64:655–666

75. Ramanan MV, Yuvarajan D (2016) Emission analysis on the influence of magnetite nanofluid on methyl ester in diesel engine. Atmosp Pollut Res 7:477–481
76. Ramarao K, Rao CJ, Sreeramulu D (2015) The experimental investigation on performance and emission characteristics of a single cylinder diesel engine using nano additives in diesel and biodiesel. Indian J Sci Technol 8(29):75718
77. Ramesh DK, Dhananjaya KJL, Hemanth KSG, Namith V, Parashuram BJ, Sharath S (2018) Study on effects of alumina nanoparticles as additive with poultry litter biodiesel on performance, combustion and emission characteristic of diesel engine. Mater Today Proc 1(1):1114–1120
78. Rani AS, Ajaad D, Karteek P, Naidu AL (2017) Experimental investigation on the performance and emission characteristics of di-diesel engine using diesel- ethanol blends and aluminium oxide nano particles. Int J Mech Product Eng Res Dev 7(5):301–310
79. Rani GJ, Rao YVH, Balakrishna B (2018) Investigation of fuel injection pressure impact on CI engine performance and emissions using biodiesel blend with alumina nano additives. Int J Mech Eng Technol 9(6):922–928
80. Ranjan A, Dawn SS, Jayaprabakar J, Nirmala N, Saikiran K, Sriram SS (2018) Experimental investigation on effect of MgO nanoparticles on cold flow properties, performance, emission and combustion characteristics of waste cooking oil biodiesel. Fuel 220:780–791
81. Rao MS, Anand RB (2016) Performance and emission characteristics improvement studies on a biodiesel fuelled DICI engine using water and AlO(OH) nanoparticles. Appl Therm Eng 98:636–645
82. Rao DS, Dash RK (2015) Effect of nanomaterials sizes on the dispersion stability of biodiesel based nanofluids. Adv Mater Lett 6:247–251
83. Sabarish R, Mahankumar D, Kumar MPJ, Manavalan S (2018) Experimental study of nano additive with biodiesel and its blends for diesel engine. Int J Pure Appl Math 118(18):967–979
84. Sahoo RR, Jain A (2019) Experimental analysis of nanofuel additives with magnetic fuel conditioning for diesel engine performance and emissions. Fuel 236:365–372
85. Sajith V, Sobhan CB, Peterson GP (2010) Experimental investigations on the effects of cerium oxide nanoparticle fuel additives on biodiesel. Adv Mech Eng 581407:1–6
86. Sandeep K, Rajashekhar CR, Karthik SR (2018) Experimental studies on effect of nano particle blended biodiesel combustion on performance and emission of CI engine. Mater Sci Eng 376:012019
87. Saraee HS, Jafarmadar S, Taghavifar H, Ashrafi SJ (2015) Reduction of emissions and fuel consumption in a compression ignition engine using nanoparticles. Int J Environ Sci Technol 12:2245–2252
88. Saraee HS, Taghavifar H, Jafarmadar S (2017) Experimental and numerical consideration of the effect of CeO_2 nanoparticles on diesel engine performance and exhaust emission with the aid of artificial neural network. Appl Therm Eng 113:663–672
89. Saxena V, Kumar N, Saxena VK (2017) A comprehensive review on combustion and stability aspects of metal nanoparticles and its additive effect on diesel and biodiesel fueled CI engine. Renew Sustain Energy Rev 70:563–588
90. Seela CR, Ravisankar B, Raju BMVA (2017) A GRNN based frame work to test the influence of nano zinc additive biodiesel blends on CI engine performance and emissions. Egypt J Petrol 27(4):641–647
91. Shaafi T, Velraj R (2015) Influence of alumina nanoparticles, ethanol and isopropanol blend as additive with diesel-soybean biodiesel blend fuel: combustion, engine performance and emissions. Renew Energy 80:655–663
92. Sharma SKS, Das RK, Sharma A (2016) Improvement in the performance and emission characteristics of diesel engine fueled with jatropha methyl ester and tyre pyrolysis oil by addition of nano additives. J Braz Soc Mech Sci Eng 38:1907–1920
93. Sivakumar M, Sundaram NS, Kumar RR, Thasthagir MHS (2018) Effect of aluminium oxide nanoparticles blended pongamia methyl ester on performance, combustion and emission characteristics of diesel engine. Renew Energy 116:518–526

94. Soudagar MEM, Nik-Ghazali N-N, Kalam MdA, Badruddin IA, Banapurmath NR, Akram N (2018) The effect of nano-additives in diesel-biodiesel fuel blends: a comprehensive review on stability, engine performance and emission characteristics. Energy Convers Manag 178:146–177
95. Srinidhi C, Madhusudhan A (2017) A diesel engine performance investigation fuelled with nickel oxide nano fuel-methyl ester. Int J Renew Energy Res 7(2):676–681
96. Sundararajan NK, Ammal ARB (2018) Improvement studies on emission and combustion characteristics of DICI engine fuelled with colloidal emulsion of diesel distillate of plastic oil, TiO_2 nanoparticles and water. Environ Sci Pollut Res 25:11595–11613
97. Vairamuthu G, Sundarapandian S, Kailasasanathan C, Thangagiri B (2016) Experimental investigation on the effects of cerium oxide nanoparticle on *Calophyllum inophyllum* (Punnai) biodiesel blended with diesel fuel in DI diesel engine modified by nozzle geometry. J Energy Inst 89:668–682
98. Vellaiyan S (2019) Enhancement in combustion, performance, and emission characteristics of a diesel engine fueled with diesel, biodiesel, and its blends by using nanoadditive. Environ Sci Pollut Res 26(10):9561–9573
99. Vellaiyan S, Amirthagadeswaran KS (2016) The role of water-in-diesel emulsion and its additives on diesel engine performance and emission levels: a retrospective review. Alex Eng J 55:2463–2472
100. Venu H, Madhavan V (2016) Effect of Al_2O_3 nanoparticles in biodiesel-diesel-ethanol blends at various injection strategies: Performance, combustion and emission characteristics. Fuel 186:176–189
101. Vinukumar K, Azhagurajan A, Vettivel SC, Vedaraman N, Lenin AH (2018) Biodiesel with nano additives from coconut shell for decreasing emissions in diesel engines. Fuel 222:180–184
102. Vinukumar K, Azhagurajan A, Vettivel SC, Vedaraman N, Lenin AH (2018) Rice husk as nanoadditive in diesel-biodiesel fuel blends used in diesel engine. J Therm Anal Calorim 131:1333–1343
103. Wu Q, Xie X, Wang Y, Roskilly T (2018) Effect of carbon coated aluminum nanoparticles as additive to biodiesel-diesel blends on performance and emission characteristics of diesel engine. Appl Energy 221:597–604
104. Yuvarajan D, Babu MD, Kumar NB, Kishore PA (2018) Experimental investigation on the influence of titanium dioxide nanofluid on emission pattern of biodiesel in a diesel engine. Atmosp Pollut Res 9:47–52

Chapter 5
Nanofuel Usage in Diesel Engines

Nanofuel application in diesel engines has many aspects and necessitates thorough investigations in order to get the answers to the following important questions [25, 37, 45, 50, 65, 67, 96, 116, 119, 126, 133]:

- What are the most important NMs activities within a diesel engine?
- What happens with the NMs at any stage of their existence from the fuel tank to exhaust?
- What are the effects of nanofuel usage on diesel engine characteristics?
- What are the effects of NMs on various parts of a diesel engine and related systems?

Within this chapter, the most important NMs activities and their role within a diesel engine are discussed in Sect. 5.1. Meanwhile, the transformations of NMs during the combustion process are addressed in Sect. 5.2. Section 5.3 discusses the effects of various NMs on engine parts and subsystems. Furthermore, NMs effects on various engine characteristics are addressed in Sects. 5.4 and 5.5. Finally, Sect. 5.6 deals with the identification of best-performing nanofuels. It has to be pointed out that this chapter deals with laboratory experimental results obtained on a diesel engine run with nanofuels and is therefore mainly focused on possible short-term benefits of NMs engagement.

5.1 Nanomaterial Activities

Within a diesel engine, NMs are heavily engaged in various mechanical, chemical, and thermal processes [15, 49, 55, 83, 95]. In this book, their engagement is roughly classified into four groups as follows:

- influences related to secondary fuel atomization due to disruption behavior and micro-explosions,

T. Kegl et al., *Nanomaterials for Environmental Application*, Green Energy and Technology, https://doi.org/10.1007/978-3-030-54708-0_5

- catalytic engagement in chemical reactions (mainly during the combustion process),
- reactivity engagement in chemical reactions, and
- thermal engagement mainly due to NMs acting as a heat sink.

5.1.1 Secondary Atomization

During the injection and combustion processes, NMs may promote secondary atomization mainly due to disruption behavior and micro-explosions (Fig. 5.1). In general, this leads to better fuel/air mixing, more rapid evaporation, and a more efficient and complete combustion process [17, 63, 83, 94, 95, 126].

During fuel delivery into the combustion chamber at high temperatures and pressures, the fuel droplets are injected into a strongly accelerating flow field with a high momentum, which results in primary fuel atomization. The presence of NMs within fuel droplets may result in increased aerodynamic shear between the droplets and the flow field. This can induce severe droplet shape deformations, which may eventually cause secondary atomization of the nanofuel droplets [16]. This disruptive behavior can vary from minor to major, depending on the density, structure, and porosity of NMs, and on thermo-physical properties of the base fuel. Besides this, droplet deformation mechanics is strongly influenced by NMs concentration and stability of nanofuels [17]. It has to be noted that disruption behavior of NMs is the only explanation for improved secondary atomization in the case when the base fuel and all of its additive components exhibit similar boiling points and volatilities. In this context, Prabu [94, 95] reported that the addition of Al_2O_3 and CeO_2 NPs mixture to *j*B100 and to D/*j*B100 base fuels improves fuel atomization due to secondary atomization. Similar behavior was also observed, for example, with CuO-D100 [109], ZnO-*ci*B100 [80], C-D/*h*B20 [113], (SiO_2 and TiO_2)-D/*bba*B20 [59], MgO-D/*wco*B20 [106], and C-D/*ts*B20 [100] nanofuels.

Apart from disruption behavior, micro-explosions are another factor that may significantly boost secondary atomization. These phenomena can be observed in:

- fuels consisting of multiple components, which exhibit different boiling points (especially various emulsions and fuel blends with alcohols) and volatilities,

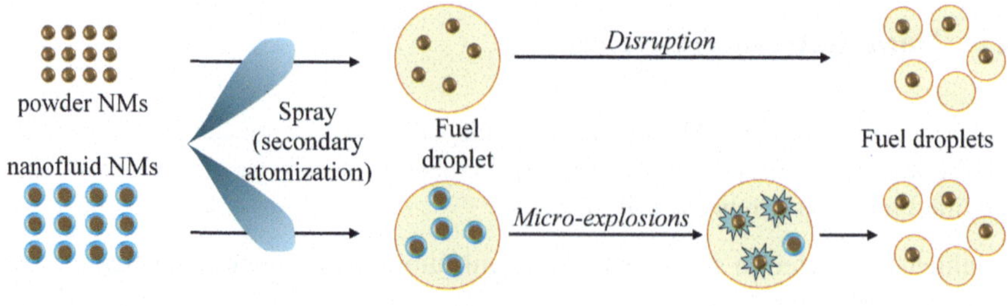

Fig. 5.1 NMs promoted secondary atomization

- fuels with added surfactants, and
- fuels with added NMs in the form of nanofluids.

In general, micro-explosions are caused by heterogeneous nucleation, where nucleation typically occurs at the droplet [117]. Namely, when droplets with components exhibiting different boiling points are exposed to high in-cylinder pressure and temperature, some components (dispersed phase) reach their boiling point faster than other components (continuous phase) [14]. This leads to explosions of lower-boiling-point components within the surrounding component layers. Such micro-explosions break the fuel droplets, resulting in a number of secondary fuel droplets of very fine size.

The role of micro-explosions, when using various fuel blends or emulsions, has been extensively discussed in the literature [4, 8, 9, 11, 12, 29, 51, 112, 117, 138, 143, 144]. Sazhin et al. [117] concluded that micro-explosions in water-in-diesel emulsion are influenced by water concentration, surfactant type, and dispersed water droplet size and temperature. Samec et al. [112] investigated the physical effects of water presence in water-in-diesel emulsion during combustion process in a diesel engine, equipped with a MC-FIS. The results indicate that water, contained within fuel droplets, undergoes spontaneous nucleation of steam bubbles at a temperature well above the vaporization temperature, depending on in-cylinder pressure. This leads to a violent conversion of water droplets into steam and fragments surrounding diesel fuel into a vast number of smaller droplets. Ismael et al. [51] investigated micro-explosions by using water-in-diesel emulsion in a diesel engine, equipped with CR-FIS. They concluded that the size of dispersed emulsion droplets is reduced significantly after the emulsion is compressed by the high-pressure fuel pump, and again after being injected through the nozzle's orifices. Additionally, droplet sizes of emulsion fuel, returned from the pump and injector to the fuel tank, are also smaller than the droplet size of initial emulsion fuel. This indicates that the physical and calorific properties of the water-in-diesel emulsion in the fuel tank can change significantly with time. Consequently, the diesel engine characteristics may vary with engine running time. Zhang et al. [143] showed that the intensity of micro-explosion obeys the parabolic rule when changing volume fraction of the hydrous ethanol in a blend of ethanol and mineral diesel at diesel engine-like conditions. Wang et al. [138] compared water-in-diesel emulsion with a blend of ethanol and mineral diesel. They concluded that micro-explosions within the blend of ethanol and mineral diesel occur at lower temperature than within water-in-diesel emulsion. It was observed that within a blend of ethanol and mineral diesel micro-explosions are predominant under evaporation conditions; therefore, micro-explosions mainly affect the evaporation characteristics. Meanwhile, within water-in-diesel emulsion micro-explosions are dominant under combustion conditions and mainly affect the combustion characteristics. On the basis of various other experimental results, it can be concluded that diesel–biodiesel–ethanol blends [11, 12] and butanol–biodiesel blends [143] also have a potential to undergo micro-explosion during evaporation and combustion processes within a diesel engine.

So far, several investigations have shown that micro-explosions may be enhanced significantly when using NMs as fuel additives [8, 9, 14, 19, 25, 29, 36, 43]. Namely, the nanosize and the high surface area-to-volume ratio of NMs may induce swollen and superheated fuel droplets, forming an extreme vaporization effect. This leads to an amplified micro-explosion effect, which promotes fuel secondary atomization into ultrafine droplets and shortens the ignition delay significantly [14, 15, 55]. Furthermore, some reports show that NMs, especially metallic NPs in the form of a nanofluid, may lead to micro-explosions in the emulsified fuel droplets due to the presence of water or alcohols within NPs [8, 9, 63]. In this context, nonmetallic NMs, in the form of a nanofluid, may also prove to be quite useful in promoting micro-explosions [8, 9, 13, 15, 19, 29, 80]. Besides the already mentioned reasons for micro-explosions, Ooi et al. [83] reported that the high micro-explosion frequency of nanofuel droplets of mineral diesel with graphite oxide (G'O) NPs can be explained also by the energetic decomposition of G'O NPs at elevated temperatures.

5.1.2 Catalytic Activity

One of the important roles of NMs is to act as a catalyst in various reactions during the combustion process and after-treatment of combustion products (Fig. 5.2) [3, 8, 9, 47, 78, 100, 125, 126]. Soudagar et al. [126] reported that various NMs can act as a catalyst to increase the rate of chemical reactions and consequently lead to complete combustion of various fuels in a diesel engine. The catalytic role of NMs is to a great extent a consequence of the large surface area-to-volume ratio, leading the fuel burning process toward the stoichiometric chemical reaction [3, 94, 95]. Namely, practically all NMs possess reactive surfaces that promote chemical reactivity and contribute to their potential to act as a catalyst [2, 13, 14, 20]. For example, it was shown that TiO_2 NPs added to the *ci*B100 base fuel accelerate the fuel combustion process. Namely, the highly reactive surface of TiO_2 NPs helps to split the hydrogen atoms from water molecules present in the fuel and the oxidation of these hydrogen atoms also takes place within the diesel engine combustion chamber [9, 126]. Besides the large surface area-to-volume ratio of NMs, size and concentration of NMs, and other NMs properties as well as the in-cylinder temperature influence their catalytic ability [7]. For example, El-Seesy et al. [33, 35] reported that the catalytic effects of G NPTs in the D/*j*B20 fuel decrease with higher engine load. Aalam et al. [2] reported that NMs addition to fuel can increase the diffusion-controlled combustion. Furthermore, Aghbashlo et al. [3] noted that the catalytic activity of metal-based NMs can improve the positive effects of micro-explosions, discussed in Sect. 5.1.1. Mingle and Lauterbach [74] and Singh and Madras [124] reported that Co_3O_4 NPs exhibit high catalytic activity for CO oxidation.

NMs acting as catalyst can also reduce the needed temperature for carbon activation. Sivakumar et al. [125] reported that Al_2O_3 NPs in the D/*pp*B25 fuel act as catalyst to accelerate the flame propagation inside the cylinder, which reduces the carbon activation temperature and promotes further burning. Logothetidis et al. [71]

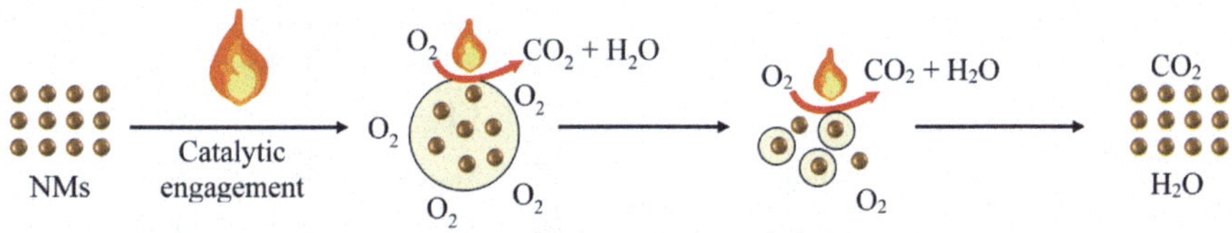

Fig. 5.2 NMs catalytic activity

reported that the size reduction of CeO_2 NPs, roughly from 30 to 5 nm, resulted in a decrease of carbon activation temperature from 600 to 250 °C.

Engagement of NMs-enriched fuels may also facilitate DPF regeneration by decreasing the oxidation temperature of carbonaceous particulate. Namely, by dosing very small quantities of NMs into the base fuel on board, it is possible to substantially lower the required soot ignition temperature within the DPF [40]. This is related to increased catalytic activity due to the presence of NMs in the particulates of exhaust gases, which effectively increases the catalytically active surface within the DPF [40]. Mohankumar and Senthilkumar [76] reported that a reduction of the needed temperature for soot oxidation can be achieved by NMs additives on the basis of Fe, Ce, Mn, Zn, Cu, and Pb NPs. For example, CeO_2 NPs and CuO NPs in a concentration of 60–100 ppm reduce soot ignition temperature to about 300 °C and DPF regeneration can been achieved at a temperature below 400 °C.

5.1.3 Reactive Activity

Some carbon-based NMs can burn during the combustion process in the cylinder and release heat [123, 129]. Furthermore, at certain conditions, some NMs can act as oxidation or reduction agents [2, 63] in reactions involving either:

- some fuel components or
- combustion products.

At this point, it should be noted that for the sake of convenience all of the mostly reported possible reactions are given in this section, although some of these reactions can actually be part of the catalytic activity of some NMs.

NMs can react with fuel components, especially when the base fuel contains water or alcohols. In such cases, NMs can react with water or alcohols to produce hydroxyl radicals, which results in various effects, including lowering of soot oxidation temperature [63]. Aalam et al. [2] reported that aluminum NPs can react with H_2O at temperatures ranging from 400 to 650 °C to generate hydrogen and promote combustion.

Apart from fuel components, NMs can also react with fuel combustion products generated by a diesel engine (Fig. 5.3). Typically, they can react directly with carbon in the soot and with NO, CO, and HC molecules; the engaged oxidation/reduction mechanisms of NMs may depend on various factors.

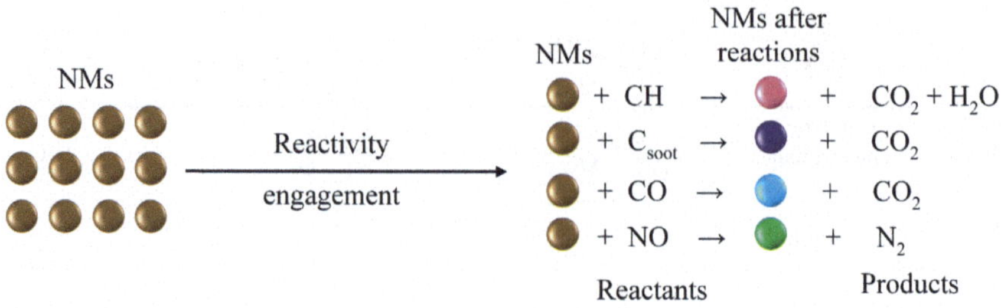

Fig. 5.3 NMs reactive activity

Among those NMs, which may react directly with combustion products under special circumstances, the most frequently addressed and investigated NMs are: CeO_2 NPs, Al_2O_3 NPs, mixture of Al_2O_3 and CeO_2 NPs, ZnO NPs, and TiO_2 NPs. The possible reactions, reported most frequently in the literature, are briefly described in the following.

CeO_2 NPs can be a chemically active component, acting as an oxygen store by releasing oxygen in the presence of reductive gases and removing oxygen by interaction with oxidizing species [42]. CeO_2 NPs can supply oxygen due to the ability of Ce to undergo a transformation from the stoichiometric valence state Ce^{+4} in CeO_2 to valence state Ce^{+3} in Ce_2O_3 via a relatively low energy reaction. Cerium oxide may provide oxygen for the oxidation of CO, HC, and soot [42, 75, 131]:

$$4CeO_2 \rightarrow 2\,Ce_2O_3 + O_2.$$

Furthermore, CeO_2 NPs can react directly with carbon atoms in soot [55, 64, 75, 110, 111]:

$$4CeO_2 + C_{soot} \rightarrow 2Ce_2O_3 + CO_2.$$

Finally, CeO_2 NPs can also act as an oxygen buffer, provide oxygen for the oxidation of CO, converting CO–CO_2 [7, 29], and react with HC to form Ce_2O_3, CO_2, and water vapor [7, 29, 75, 111]:

$$2CeO_2 + CO \rightarrow Ce_2O_3 + CO_2$$
$$(4x + 2y)CeO_2 + C_xH_{2y} \rightarrow (2x + y)Ce_2O_3 + xCO_2 + yH_2O.$$

Ce_2O_3, formed by the oxidation of soot, carbon monoxide, and hydrocarbons, can get re-oxidized to CeO_2 through the reduction of NO [7, 29, 55, 75, 110, 111, 115]:

$$Ce_2O_3 + NO \rightarrow 2CeO_2 + \frac{1}{2}N_2.$$

Al_2O_3 NPs in the form of a nanofluid can generate additional oxygen, which reacts with C atoms [19]. Furthermore, Al_2O_3 NPs can react with hydroxyl radicals to form AlO_2, which can reduce unburned HC to form Al_2O_3, CO_2, and water vapor [95]:

$$\begin{aligned}
&Al_2O_3 + 2OH \rightarrow 2AlO_2 + H_2O \\
&(4x + 2y)AlO_2 + C_xH_{2y} \rightarrow (2x + y)Al_2O_3 + xCO_2 + yH_2O.
\end{aligned}$$

Venu et al. [135] reported that at higher in-cylinder temperatures, Al_2O_3 NPs react with NO to form AlO_2 and N_2, thereby lowering the possibility of NO_x formation:

$$Al_2O_3 + NO \rightarrow 2AlO_2 + \frac{1}{2}N_2.$$

The reaction of mixture of Al_2O_3 and CeO_2 NPs in a diesel combustion chamber can convert the formed NO into N_2 and the formed CO into CO_2 [94, 95]:

$$\begin{aligned}
&Ce_2O_3 + Al_2O_3 \rightarrow 2CeAlO_3 \\
&4CeAlO_3 + 2NO \rightarrow 4CeO_2 + 2Al_2O_3 + N_2 \\
&CeAlO_3 + CO \rightarrow CeO_2 + CO_2 + Al.
\end{aligned}$$

Co_3O_4 NPs can donate surface lattice oxygen at temperatures over 1000 °C to form a stable cobaltous oxide. Co_3O_4 NPs may also provide oxygen for the oxidation of CO, HC, and soot, or may absorb oxygen for the reduction of NO_x throughout the diesel engine combustion process [108]:

$$Co_3O_4 \rightarrow 3CoO + \frac{1}{2}O_2.$$

ZnO NPs and TiO_2 NPs have been found capable of enabling hydroxyl radical reactions with unburned HC and further oxidizing HC into H_2O and CO_2 [8, 9]. ZnO NPs can also act as an oxygen buffer at higher in-cylinder temperatures and release oxygen which oxidizes the major pollutant CO into CO_2 [120]:

$$ZnO + CO \rightarrow Zn + CO_2.$$

Furthermore, Sharma et al. in [122] reported that ZnO NPs exhibit features of high adsorption of sulfur, high activity, and high mechanical strength. They can maintain very high activity and sulfur adsorption capacity also at very high temperatures and harsh conditions found within a diesel engine. Thus, ZnO NPs can be engaged to react with the toxic compounds that are formed during the combustion process:

$$\begin{aligned}
&ZnO + H_2S \rightarrow ZnS + H_2O \\
&2ZnO + CS_2 \rightarrow 2ZnS + CO_2.
\end{aligned}$$

At in-cylinder temperature of about 1500 °C, a considerable amount of pollutants can get reduced to comparatively nonharmful products due to the ZnO reactivity.

5.1.4 Thermal Activity

Some NMs, especially in the form of a nanofluid, act in a diesel combustion chamber as a heat sink by absorbing the heat and reducing the in-cylinder temperature (Fig. 5.4). In general, this thermal activity of NMs typically results in a reduced NO_x emission [19, 78]. Ooi et al. [83] reported that the heat absorbed by the NMs is transported away from the droplets to the fuel vapor/air mixing region by diffusion. This results in droplet cooling, while more heat is supplied by the NMs at the fuel vapor/air mixing region to be ignited earlier.

Annamalai et al. [7] reported that the CeO_2 NPs added to the *lg*O/W5 fuel absorb the heat and reduce the in-cylinder temperature due to high thermal conductivity of this nanofuel. Another reason for heat absorption is explained by Ashok et al. in [8, 9] by using ZnO-*ci*B100 and TiO_2-*ci*B100 nanofuels. Here, the formation of peroxide and peroxyl radicals takes place during the diesel combustion process; these products are then further converted into hydroxyl radicals by absorbing heat. By the addition of C NTs in the form of a nanofluid to the *j*B/W5 fuel, the heat sink effect was also observed due to water droplets vaporizing during the combustion, thereby decreasing notably the burning in-cylinder temperature [13, 15].

5.2 Transformation of Nanomaterials

A rather modest extent of investigations has been conducted so far on the transformation of NMs on their path from the FIS to exhaust system end of a diesel engine. On this path, the NMs are exposed to very high temperatures and pressures for a few milliseconds during the combustion process as well as to moderate temperatures and pressures for tenths of a second while passing the exhaust system. The research results indicate that NMs can transform significantly during this travel [1, 24, 76].

Some investigation results reveal that some oxide NPs may dissociate at high temperature during the combustion process in the cylinder [1, 126]. For example,

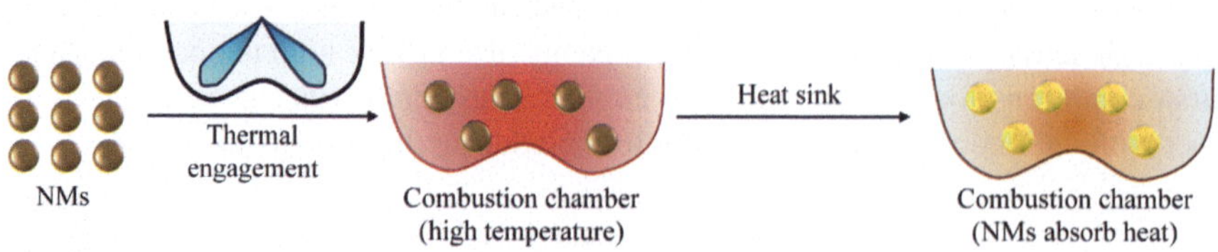

Fig. 5.4 NMs thermal activity

Soudagar et al. [126], Aalam and Saravanan [1], and Venu et al. [135] reported that Al_2O_3 NPs are unstable at high temperature and decompose as:

$$Al_2O_3 \rightarrow Al_2O + O_2 \quad \text{and} \quad Al_2O \rightarrow 2Al + \frac{1}{2}O_2.$$

Co_3O_4 NPs can form a stable CoO at temperatures over 1000 °C [108] as follows:

$$Co_3O_4 \rightarrow 3CoO + \frac{1}{2}O_2.$$

Furthermore, CeO_2 NPs can be converted to Ce_2O_3 due to the ability of Ce to undergo a transformation from the stoichiometric valence state Ce^{+4} in CeO_2 to valence state Ce^{+3} in Ce_2O_3 as:

$$4CeO_2 \rightarrow 2Ce_2O_3 + O_2$$

via a relatively low energy reaction [42]. Ce_2O_3 is an unstable compound and transforms back to CeO_2 as:

$$Ce_2O_3 + \frac{1}{2}O_2 \rightarrow 2CeO_2$$

when excess oxygen becomes available in the exhaust gas [42, 76]. These reactions might cause the impression that CeO_2 NPs will exit the exhaust system more or less untouched. However, as some investigations confirmed, the CeO_2 NPs might still transform quite substantially [18, 24]. Batley et al. [18] investigated captured exhaust particulates removed from the filter after combustion in a diesel engine with and without added CeO_2 NPs to the D100 base fuel. The measurements revealed that CeO_2 NPs in the range of 6–14 nm were transformed during combustion into larger particles of approximately 43 nm. Apart from increased dimensions, however, changes in the unit cell dimensions or crystalline form were not observed. Furthermore, Dale et al. [24] reported that CeO_2 NPs transformed from crystals of 5–7 nm size in the nanofuel to near-spherical 50–300-nm single crystals in the diesel exhaust system. Moreover, the surfactant coating present on CeO_2 NPs was removed during the combustion process. Instead, a new carbonaceous 2–5 nm thick coating formed on approximately 30% of the NPs in the exhaust system. Dale et al. [24] showed also that morphological alterations change the termination surfaces of the crystal, while a restructuring of the internal crystallinity of the NPs leads to superlattice formation in some of the crystals [24].

Wu et al. [139] investigated the phase structure change of Al@C NPs after combustion in a diesel engine run with the Al@C-D/*p*B10/E4 nanofuel. It was found out that before combustion the Al@C NPs were quasi-spherical with sizes of 60–120 nm, having a disordered outer layer with thicknesses around 5 nm, indicating amorphous structure of graphite, and aluminum core, indicating a crystallinity structure [139]. After the combustion in the engine, the recovered particulates exhibited diameters of about 160 nm and the core@shell structure of the NPs was not very clear.

Zhang and Balasubramanian [141] reported that addition of iron NPs to diesel fuel mainly results in emission of Fe in its elemental form. Only a very small amount was found in the oxide form.

Under the bottom line, one can say that the properties on NMs may change significantly during their lifetime in a diesel engine due to already known as well as yet unknown transformations. This surely makes more difficult to predict their effects on the environment and human health [24].

5.3 Effects on Diesel Engine Parts and Systems

NMs present in nanofuels may also have some effects on various engine parts and systems. Although there are not many results on long-term diesel engine testing, a brief outline of available data is given in the following sections.

As it looks, NMs can potentially have an important role in preventing carbon, metals, and other materials to deposit on cylinder walls, which could significantly decrease friction [63, 64, 75]. Mirzajanzadeh et al. [75], Annamalai et al. [7], and Khalife et al. [64] explained that CeO_2 NPs, dispersed in the fuel, produce millions of nanoclusters whose explosions result in the disintegration and decomposition of the deposits. In this way, CeO_2 NPs prevent the precipitation of nonpolar components on the cylinder walls. Furthermore, Sivakumar et al. [125] reported that addition of Al_2O_3 NPs to the D/*pp*B25 base fuel reduces the formation of carbon deposits which in turn leads to decreased friction power of the cylinder. Aalam et al. [2] reported that Al_2O_3 NPs in the D/*zj*B25 fuel can also oxidize the carbon deposits in the combustion chamber.

As can be seen from Tables 4.14–4.25, the addition of NMs to base fuels may also alter the viscosity of the fuel. In general, it is known that high fuel viscosity tends to form engine deposits [69]. Moreover, long-term durability problems may be related to the use of high viscosity fuels, and carbon deposits may dilute in the engine lubricant, leading to engine damage. Also, high viscosity fuels require more energy in the fuel pump and increase wear in the FIS. At low temperatures, fuel filter plugging problems may occur since viscosity increases with decreasing temperatures. On the contrary, fuels with low viscosity may not provide sufficient lubrication for the FIS, leading to greater pump and injector leakage, reducing maximum fuel delivery, and thus decreasing power output [69, 116]. Since NMs may alter the viscosity of fuel, their usage should be carefully investigated also in this context, especially if high viscosity nanofuels are planned to be used at low temperatures.

The altered physical and chemical properties of NMs-enriched fuels require a modified nozzle geometry, a modified piston geometry in order to ensure an adequate swirl in the combustion chamber, and modified programming/maps of the ECU. Dhinesh et al. [30] tested various piston geometries when using 20 ppm of CeO_2 NPs added to the D/B20 base fuel. They concluded that among the tested hemispherical combustion chamber, shallow depth reentrant combustion bowl, toroidal reentrant combustion bowl, and toroidal combustion bowl, the toroidal reentrant combustion bowl gives the best engine characteristics for the tested Kirloskar TV1 diesel engine.

Some authors reported that during their experimental work of duration up to 2 months no changes on the diesel engine components were observed [33, 35]. However, they noted that long-term investigations on the life of the engine exhaust manifold and FIS would be needed to determine such possible effects in the long term.

5.4 Effects on Diesel Engine Process Characteristics

5.4.1 Injection Process Characteristics

According to literature, the addition of NMs to base fuels may influence injection timing and injection pressure. Under special circumstances, even a blocking of the injection process may become possible. These effects might be observed in an engine equipped with MC-FIS.

Injection timing. In a MC-FIS, the injection timing is controlled indirectly by many parameters, including the speed of pressure wave propagation, which in turn greatly depends on fuel density. Since the density of a nanofuel is typically higher than the one of the corresponding base fuels, the injection timing is often advanced when some base fuel is replaced by a nanofuel. Namely, higher fuel density is strongly related to higher bulk modulus and sound velocity. Especially in a MC-FIS, this is directly related to the advanced injection timing [43, 62, 125]. For example, Srinidhi et al. [128] investigated the influence on static injection timing in a MC-FIS when NiO NPs are added to the D/*n*B25 base fuel. It was found out that the presence of 100 ppm of NiO NPs notably advances the injection timing with the potential to improve brake thermal efficiency and reduce almost all harmful emissions (except NO_x) of a diesel engine.

Injection pressure. When the viscosity of a fuel is increased, this is typically reflected in a higher injection pressure within the MC-FIS [62, 116]. Consequently, since NMs addition often increases the viscosity of the base fuel significantly [33, 38, 68, 70, 77, 91, 92, 98, 105], higher injection pressure in the MC-FIS can typically be expected [116].

Block of injection process. Örs et al. [85] warned that TiO_2 NPs addition to the D/*wco*B20 base fuel raises the chances for the injection process to be blocked. This is because the NPs may exhibit a tendency to clump due to their large surface area and surface activity.

Fuel property variations induced by NMs addition may obviously influence the injection characteristics within the MC-FIS significantly. By keeping this in mind, it is obvious that for a CR-FIS a reprogramming of the injection control parameters (injection pressure, injection timing, and injection rate) will be necessary in order to obtain the desired fuel spray and combustion process as well as engine characteristics.

5.4.2 *Fuel Spray Characteristics*

To analyze the effects of NMs addition on fuel spray characteristics, the NMs influence needs to be discussed with respect to:

- macroscopic and
- microscopic fuel spray characteristics.

Macroscopic fuel spray characteristics. Ahmad and Hossain [5] tested Al_2O_3 NPs added in the form of a nanofluid to the *wco*B/Bu10 base fuel. They reported that this improves fuel spray characteristics by widening the fuel spray angle and improving spray dispersion which leads to better atomization as more fuel particles get in contact with air. Kannaiyan et al. [58] tested the sprays of the GTL100 base fuel and Al_2O_3-GTL100 nanofuel. The GTL100 spray formed a steady-operation cone angle of $85 \pm 2°$ after about 25 ms, whereas the nanofuel spray reached its steady operation somewhat earlier. Kannaiyan and Sadr [57] reported that with the increase of Al_2O_3 NPs dosage in the GTL100 base fuel the liquid spray breakup length decreases. This change might be caused by reduced surface tension and increased fuel viscosity as a result of increased dosage of Al_2O_3 NPs. Furthermore, the perforations in the liquid spray were observed closer to the nozzle exit as the Al_2O_3 NPs dosage in GTL100 was increased [57, 58]. Saraee et al. [114] concluded that Ag NPs additive to the D100 base fuel can widen the spray cone angle and facilitate fuel dispersion in the combustion chamber due to differences in density and momentum in the injection period and due to the interaction of Ag NPs with the base fuel.

Microscopic fuel spray characteristics. Nanofuels typically exhibit smaller mean droplet diameters within the spray than those obtained with the base fuel. Tests with the Al_2O_3-GTL100 nanofuel confirmed this in most of the radial locations, where shorter liquid spray breakup lengths provided longer distances/duration for secondary atomization [57]. Furthermore, Kannaiyan and Sadr [57] concluded that in the presence of Al_2O_3 NPs, the spray characteristics within the near-nozzle regions are more affected than in the downstream regions. For example, Kannaiyan et al. [58] reported that the liquid spray dynamics of the GTL100 base fuel is notably altered near the nozzle exit, when adding Al_2O_3 NPs. El-Seesy et al. [34] reported that by increasing the amount of GO NPs, added in the form of a nanofluid to *j*B100, the kinematic viscosity of the GO-*j*B100 nanofuel decreases. Meanwhile, its cetane number increases, which has a positive effect on fuel spray atomization. The positive effect of low kinematic viscosity on enhanced fuel atomization was also reported by El-Seesy and Hassan [31] from tests with mwC NTs and G NPTs added to the *j*B/Bu40 base fuel. Örs et al. [85] tested TiO_2-D/*wco*B20 and TiO_2-D/*wco*B20/Bu10 nanofuels. They reported that, compared to base fuels, better atomization of the nanofuels was observed due to higher evaporation rate, which is a consequence of high surface area of the NPs. Prabu [95] reported that the added mixture of Al_2O_3 and CeO_2 NPs into the *j*B100 and D/*j*B20 base fuels promotes the fuel droplets to become swollen and superheated. This forms an extreme vaporization effect, which leads to secondary atomization. Saraee et al. [114] tested Ag NPs added to the D100 base fuel. They concluded that the addition of Ag NPs can reduce the mean fuel droplet diameters due to effect of reduced fuel cohesion, caused by these NPs.

5.4.3 Combustion Process Characteristics

NMs addition obviously affects all of the most important combustion characteristics: ignition delay (φ_{id}), maximal values of in-cylinder pressure $\left(p_{\mathrm{cyl,max}}\right)$, heat release rate $\left(\dot{q}_{\mathrm{h,max}}\right)$, and exhaust gas temperature $\left(T_{\mathrm{eg,max}}\right)$.

Ignition delay. The experimentally obtained results of practically all recent investigations of nanofuels, collected in Tables 4.14–4.25, confirm that NMs fuel additives typically reduce φ_{id}. The most frequently reported reason for φ_{id} reduction is the higher CN of the nanofuels when compared to the corresponding base fuels [7–9, 13, 15, 33–35, 84, 125, 130]. Furthermore, φ_{id} can also be reduced because of the catalytic activity of NMs, which helps to improve the combustion of the fuel/air mixture [6, 34, 94, 95, 125].

In-cylinder pressure. Maximal in-cylinder pressure $p_{\mathrm{cyl,max}}$ can either increase or decrease when NMs are engaged as fuel additives. The main reasons for $p_{\mathrm{cyl,max}}$ increase are higher CN [6, 34, 80, 100], lower viscosity [1], and higher thermal conductivity [33] of the nanofuels. Besides this, some authors explain the increase of the $p_{\mathrm{cyl,max}}$ by enhancements in the combustion process due to the catalytic activity of NMs [35, 82, 85, 94, 95, 100, 125] and reactive activity of NMs, especially, related to their role as oxygen buffer [8, 97]. When the NMs are added to the base fuel in the form of a nanofluid, $p_{\mathrm{cyl,max}}$ can be higher due to the combined effects of NPs and water particles present in the fuel [34, 80]. Some authors reported a decrease of $p_{\mathrm{cyl,max}}$ due to NMs addition on the account of evaporation and atomization rate [13, 15, 31, 60, 73, 103]. By using some carbon NMs-enriched fuels, $p_{\mathrm{cyl,max}}$ can also be decreased and this is typically accompanied with a maximum value shift closer to the TDC [13, 31].

Heat release rate. NMs addition may influence $\dot{q}_{\mathrm{h,max}}$ in various ways. The main reasons for a considerable increase of $\dot{q}_{\mathrm{h,max}}$ are the catalytic activity of NMs [1, 6, 8, 9, 27, 28, 35, 85, 92, 94, 95, 97, 100], higher calorific value H_{f} [2], and higher thermal conductivity [34] of the nanofuels. When the NMs are added to the base fuel in the form of a nanofluid, a significant reduction of $\dot{q}_{\mathrm{h,max}}$ can usually be observed due to the heat sink effect [13, 15].

Exhaust gas temperature. The impact of NMs addition on $T_{\mathrm{eg,max}}$ is pretty much variable. The main reason for increased $T_{\mathrm{eg,max}}$ is the higher H_{f} of the nanofuel when compared to the base fuel [21, 73, 92]. Chen et al. [21] investigated the addition of Al_2O_3 NPs, C NTs, and SiO_2 NPs to the D100 base fuel on a Yanmar TF120 diesel engine. They observed that the highest $T_{\mathrm{eg,max}}$ was achieved with C NTs due to the highest H_{f} and the longest φ_{id}, observed for the C-D100 nanofuel. This contributes to a slower fuel burning process, and the combustion gas exits the combustion chamber at a higher temperature. On the other hand, $T_{eg,max}$ may decrease due to higher thermal diffusivity and lower viscosity of the nanofuels [54] and due to catalytic activity of NMs [6, 15]. Kumar et al. [66] reported that by adding the γ-Fe_2O_3 NPs to the D/*pp*B20 base fuel, the $\varphi_{ig.d}$ was reduced whereas the CN of the fuel increased and hence the major part of the combustion completed before the TDC; this ensured better combustion, which resulted in reduced $T_{eg,max}$.

5.5 Effects on Diesel Engine Characteristics

Nanofuel usage is being investigated with respect to a wide range of diesel engine characteristics related to engine ecology, economy, and performance. In this work, focus is given on a selection of some of the most important characteristics, which includes: NO_x, HC, CO, smoke, and PM exhaust emissions, brake-specific fuel consumption (g_e), effective power (P_e), and brake thermal efficiency (η_{th}). These characteristics are considered in dependence on the engaged NMs, base fuel, engine type, and operating condition. It is important to note that only the results of the most efficient nanofuels (optimized NMs dosage and size with respect to the base fuel) are presented. Furthermore, an additional condition for a reported result to be included into this book was that the reported result was (or could be) compared to that obtained with neat base fuel. Note that this comparison may relate either to the relative average variation (in case of several tested engine operating regimes) or to the relative specific variation (in case of a single tested engine operating regime) observed for a specific characteristic.

A glance at engine characteristics variations induced by nanofuel usage indicates that a full analysis of various influences is far from being a straightforward task. To simplify the situation, a specific combination of the used NMs, their size and dosage, tested base fuel, tested engine with specific FIS type, and the corresponding operating conditions is denoted here as a *case* and identified by a unique *case name*, as given in Tables 5.1, 5.2, 5.3, 5.4, 5.5, 5.6, 5.7, 5.8, 5.9, 5.10, 5.11, 5.12, and 5.13. Additional abbreviations used in these tables are: variable compression rate (VCR), direct injection (DI), exhaust gas recirculation (EGR), mechanically controlled fuel injection system (MC-FIS), and common rail fuel injection system (CR-FIS).

In the following sections, all of the considered nanofuels are grouped with respect to the underlying NMs type. For each group, the most relevant fuel and engine data are given in Tables 5.1, 5.2, 5.3, 5.4, 5.5, 5.6, 5.7, 5.8, 5.9, 5.10, 5.11, 5.12, and 5.13. The NMs effects on engine characteristics are given graphically by relative variations of the characteristics in Figs. 5.5, 5.6, 5.7, 5.8, 5.9, 5.10, 5.11, 5.12, 5.13, 5.14, 5.15, 5.16, 5.17, and 5.18. Additionally, the most relevant NMs effects are discussed briefly.

Table 5.1 Test cases with Ag-based NMs

Nano-additive	NMs: size dosage	Base fuel	Diesel engine/FIS; operating conditions	Case name	References
Ag NPs	30–50 nm 40 ppm	D100	Motorsazan MF-399/DI; 1800 rpm, full load	Ag-1	[114]
	80–90 nm 50 ppm	*pp*B100	Kirloskar/MC-FIS; 770 rpm, full load	Ag-2	[81]
	50 nm 120 ppm	*wco*B100	NA/MC-FIS; 1000 rpm, full load	Ag-3	[79]
Ag_2O NPs	<100 nm 10 ppm	*n*B100	Kirloskar AV1/MC-FIS; 1500 rpm, full load	Ag-4	[26]
	<100 nm 10 ppm	*p*B100	NA/MC-FIS; 1500 rpm, full load	Ag-5	[27]

5.5.1 NMs Based on Silver

The cases related to tests done with nanofuels obtained by addition of silver-based NMs to various base fuels are collected in Table 5.1. The corresponding reported variations of engine characteristics are shown in Fig. 5.5.

Ag-based NPs added to various base fuels can either increase or decrease NO_x and HC emissions, decrease CO and smoke emissions as well as specific fuel consumption g_e, and improve engine performance. Ag-based NPs can be used efficiently for the oxidation of organic compounds and also for the reduction of nitro compounds [114]. NO_x emission can be reduced due to lower in-cylinder temperature, caused by a higher rate of convective heat transfer, NMs heat sink effect, and shortened combustion duration in cases Ag-1 [114], Ag-4 [26], and Ag-5 [27]. The addition of Ag NPs to the *pp*B100 and *wco*B100 biodiesel fuels can lead to faster premixed combustion, which leads to higher in-cylinder temperature and higher NO_x emission in cases Ag-2 [81] and Ag-3 [79]. HC emission increased only in case Ag-2 [81]. In all other cases, CO and HC emissions are reduced due to Ag NPs ability to enhance air/fuel mixing and the combustion process [114], and due to Ag_2O NPs ability to provide oxygen for HC and CO oxidation reactions. In cases Ag-2, Ag-4, and Ag-5, the measured smoke emission was reduced by over 25%. Among all cases involving silver-based NPs, the most promising result was obtained by addition of Ag_2O NPs to palm biodiesel (case Ag-5) which delivers a good compromise between harmful emissions, specific fuel consumption, engine power, and thermal efficiency.

Table 5.2 Test cases with Al-based NMs

Nano-additive	NMs: size dosage	Base fuel	Diesel engine/FIS; operating conditions	Case name	References
Al_2O_3 NPs	50 nm 10 wt%	D100	Fiat TD 313/DI; 1500 rpm, full load	Al-1	[19]
	20–30 nm 100 ppm	D100	Yanmar TF120M/DI; 1800 rpm, full load	Al-2	[21]
	27–43 nm 100 ppm	D100	Lombardini 15LD 350/DI; various speeds and loads	Al-3	[41]
	≤50 nm 100 ppm	D100	Kirloskar AV1/MC-FIS; 1500 rpm, full load	Al-4	[56]
	5–150 nm 0.5 wt%	D100	Kirloskar TV1/MC-FIS; 1500 rpm, full load	Al-5	[73]
	<100 nm 0.006 wt%	D100	Kirloskar AKV-3/MC-FIS; various speeds and loads	Al-6	[78]
	34 nm 50 ppm	D/W10	Diesel engine/MC-FIS; 3000 rpm, full load	Al-7	[43]
	30–50 nm 40 ppm	D/*ci*B20	Kirloskar TV1/MC-FIS; 1500 rpm, various loads	Al-8	[6]
			Kirloskar TV1, 20% EGR/MC-FIS; 1500 rpm, various loads	Al-9	
	50 nm 100 ppm	D/*gs*B20	Kirloskar/MC-FIS; 1500 rpm, full load	Al-10	[60]
	<50 nm 0.5 g/L	D/*j*B20	Diesel engine/DI; 1500 rpm, various loads	Al-11	[90]
	20–50 nm 30 mg/L	D/*jj*B20	DEUTZ F1L511/DI; various speeds and loads	Al-12	[35]
	32–48 nm 100 ppm	D/*m*B20	Kirloskar AV1/CR DI; 1500 rpm, various loads	Al-13	[1]

(continued)

Table 5.2 (continued)

Nano-additive	NMs: size dosage	Base fuel	Diesel engine/FIS; operating conditions	Case name	References
	30–50 nm 100 ppm	D/*m*B25	Kirloskar TV1/MC-FIS; 1500 rpm, full load	Al-14	[105]
	20–50 nm 30 mg/L	D/*pl*B20	Diesel engine/MC-FIS; 1500 rpm, full load	Al-15	[103]
	<100 nm 100 ppm	D/*pp*B25	Kirloskar AV1/MC-FIS; 1500 rpm, full load	Al-16	[125]
	20 nm 60 ppm	D/*ts*B20	Kirloskar TV1/MC-FIS; 1500 rpm, full load	Al-17	[100]
	50 nm 90 ppm	D/*wco*B10	Lombardini 3LD510/CR-FIS; various speeds, full load	Al-18	[47]
	<100 nm 25 ppm	D/*zj*B25	Kirloskar AV1/CR DI; 1500 rpm, various loads	Al-19	[2]
	<100 nm 10 wt%	*cns*B100	Lombardini 3LD510/DI; 1800 rpm, full load	Al-20	[99]
	30 nm 50 ppm	D/E35	Kirloskar AVI/DI; 1500 rpm, full load	Al-21	[104]
	20 nm 50 mg/L	D/*j*B20/E10	Kirloskar TAF1/DI; 1500 rpm, various loads	Al-22	[134]
	20–50 nm 75 mg/L	D/*jj*B10/E20	HATZ-1B30-2/DI; 1500 rpm, various loads	Al-23	[82]
AlOOH NPs	15 nm 200 ppm	D/*co*B20	Kirloskar TV1/DI; 1500 rpm, full load	Al-24	[98]
	<20 nm 100 ppm	*j*B/W10	Kirloskar TAF1/DI; 1500 rpm, full load	Al-25	[107]

5.5.2 NMs Based on Aluminum

The cases involving aluminum-based NPs are collected in Table 5.2. The corresponding reported variations of engine characteristics are shown in Fig. 5.6. In the

Table 5.3 Test cases with Ce-based NMs

Nano-additive	NMs: size dosage	Base fuel	Diesel engine/FIS; operating conditions	Case name	References
CeO_2 NPs	10–30 nm 40 ppm	D100	Motorsazan-Iran MF-399/DI; various speeds, various loads	Ce-1	[115]
	50 nm 25 ppm	D100	HATZ 1B30/DI; 2000 rpm, full load	Ce-2	[84]
	16 nm 30 ppm	*lg*O/W5	Kirloskar TV1/MC-FIS; 1500 rpm, full load	Ce-3	[7]
	25 nm 40 ppm	D/*ci*B50	Kirloskar TV1/MC-FIS; 1500 rpm, full load	Ce-4	[131]
	<100 nm 50 ppm	D/*cs*B10	Diesel engine VCR ($\varepsilon = 14$)/DI; 1500 rpm, various load	Ce-5	[91]
	<100 nm 40 mg/L	D/*cs*B20	Kirloskar/DI; 3000 rpm, full load	Ce-6	[102]
	<100 nm 20 ppm	D/*lg*B25	Kirloskar TV1/MC-FIS; 1500 rpm, full load	Ce-7	[29]
	<25 nm 150 ppm	D/*m*B20	Kirloskar TV1/DI; 1500 rpm, full load	Ce-8	[68]
	<100 nm 90 ppm	D/*r*B5/W4	Marine diesel engine/MC-FIS; 1000 rpm, full load	Ce-9	[55]
	<100 nm 90 ppm	D/*wco*B5/W3	Ricardo E6/DI; 1000 rpm, full load	Ce-10	[64]
	15–30 nm 90 ppm	D/*wco*B5/W3	Ricardo E6/DI; 1000 rpm, full load	Ce-11	[3]
	25 nm 40 ppm	*ci*B100	Kirloskar TV1/MC-FIS; 1500 rpm, full load	Ce-12	[131]
	<100 nm 40 ppm	*j*B100	Diesel NA/DI; 1500 rpm, full load	Ce-13	[111]
	20–30 nm 100 ppm	*j*B/*tp*O10	Greaves Cotton 5520/MC-FIS; 3000 rpm, full load	Ce-14	[121]
	15 nm 30 ppm	*no*B/W15	Kirloskar/MC-FIS; 1500 rpm, full load	Ce-15	[29]

Table 5.4 Test cases with Co-based NMs

Nano-additive	NMs: size dosage	Base fuel	Diesel engine/FIS; operating conditions	Case name	References
Co_3O_4 NPs	<100 nm 100 ppm	D/*gs*B20	Kirloskar/MC-FIS; 1500 rpm, full load	Co-1	[61]
	30 nm 50 ppm	D/*wfo*B20	Diesel engine, urea SCR/DI; 1500 rpm, full load	Co-2	[72]
	<100 nm 150 mg/L	*ci*B100	Kirloskar/MC-FIS; 1500 rpm, full load	Co-3	[53]
	<100 nm 100 ppm	*j*B100	Kirloskar TAF-1/MC-FIS; 1500 rpm, full load	Co-4	[108]

Table 5.5 Test cases with Cu-based NMs

Nano-additive	NMs: size dosage	Base fuel	Diesel engine/FIS; operating conditions	Case name	References
CuO NPs	30–50 nm 50 ppm	D100	Lombardini 15LD 350/DI; various speeds, various loads	Cu-1	[41]
	<100 nm 0.006 wt%	D100	Kirloskar AKV-3/MC-FIS; various speeds, various loads	Cu-2	[78]
	<100 nm 0.5 wt%	D100	Kirloskar/DI; 1500 rpm, full load	Cu-3	[109]
	32 nm 50 ppm	D/W10	Diesel engine/MC-FIS; 3000 rpm, full load	Cu-4	[43]
	<100 nm 80 ppm	D/*l*B20	Kirloskar AV1/MC-FIS; 1500 rpm, full load	Cu-5	[52]
	<50 nm 50 mg/L	D/*pp*B20	Diesel engine VCR ($\varepsilon = 17.5$)/DI; 1500 rpm, full load	Cu-6	[23]
	50 nm 100 ppm	D/*pp*B20	Kirloskar AV1/MC-FIS; 1500 rpm, full load	Cu-7	[92]

cases Al-5, Al-8, Al-9, Al-14, and Al-17, the used engine, MC-FIS, and engine operating regimes are the same. Achupogu et al. [6] tested also the influence of EGR (cases Al-7 and Al-8).

Addition of Al_2O_3 NPs to mineral diesel affects positively HC, CO, smoke, and PM emissions as well as specific fuel consumption g_e, engine power P_e, and thermal

Table 5.6 Test cases with Fe-based NMs

Nano-additive	NMs: size dosage	Base fuel	Diesel engine/FIS; operating conditions	Case name	References
γ-Fe_2O_3 NPs	35 nm 100 ppm	D100	Kirloskar AV1/MC-FIS; 1500 rpm, full load	Fe-1	[56]
	30–43 nm 0.5 wt%	D100	Kirloskar TV1/MC-FIS; 1500 rpm, full load	Fe-2	[73]
	30 nm 1 wt%	D/*pp*B20	Kirloskar TV1/MC-FIS; 1500 rpm, full load	Fe-3	[66]
	<100 nm 100 ppm	D/*pp*B20	Kirloskar AV1/DI; 1500 rpm, full load	Fe-4	[77]
Fe_3O_4 NPs	10 nm 1 wt%	*rb*B100	Kirloskar AV1/DI; 1500 rpm, full load	Fe-5	[101]

Table 5.7 Test cases with Mg-based NMs

Nano-additive	NMs: size dosage	Base fuel	Diesel engine/FIS; operating conditions	Case name	References
MgO NPs	33 nm 50 ppm	D/W10	Diesel engine/MC-FIS; 3000 rpm, full load	Mg-1	[43]
	20–30 nm 30 ppm	D/*wco*B20	Kirloskar TV1/MC-FIS; 1500 rpm, various load	Mg-2	[106]
	10–30 nm 50 ppm	*r*B100	Mitsubishi Canter 4D34-2A/DI; various speeds, full load	Mg-3	[86]
	20–30 nm 30 ppm	*wco*B100	Kirloskar TV1/MC-FIS; 1500 rpm, various load	Mg-4	[106]

efficiency η_{th}. Chaichan et al. [19] investigated NO_x emission at various engine loads and speeds when using various dosages of Al_2O_3 NPs, added in the form of nanofluid to D100 fuel. At all engine speed regimes (at constant medium load) as well as at all engine loads (at constant engine speed of 1500 rpm), the NO_x emission was decreasing by increasing the dosage of Al_2O_3 NPs. This was attributed to Al_2O_3 NPs ability to act as a heat sink and the consequent reduction of the in-cylinder temperature [19]. Only in cases Al-2 and Al-5, the addition of Al_2O_3 NPs increased

Table 5.8 Test cases with Ti-based NMs

Nano-additive	NMs: size dosage	Base fuel	Diesel engine/FIS; operating conditions	Case name	References
TiO_2 NPs	10–20 nm 80 mg/L	D100	VCR NA/DI; 1500 rpm, full load	Ti-1	[22]
	10–50 nm 40 ppm	D/W5	Kirloskar TAF1/MC-FIS; 1500 rpm, full load	Ti-2	[130]
	10–20 nm 25 ppm	D/*me*B20	Kirloskar TAF1/MC-FIS; 1500 rpm, various loads	Ti-3	[70]
			Kirloskar TAF1, 20% EGR/DI; 1500 rpm, full load	Ti-4	
	10–50 nm 40 ppm	*p*O/W5	Kirloskar TAF1/MC-FIS; 1500 rpm, full load	Ti-5	[130]
	<100 nm 40 ppm	D/*ci*B20	Kirloskar/MC-FIS; 1500 rpm, full load	Ti-6	[97]
			Kirloskar, 20% EGR/MC-FIS 1500 rpm, full load	Ti-7	
	<100 nm 0.01 wt%	D/*wco*B20	3 LD 510/DI; 2800 rpm, various loads	Ti-8	[85]
	<100 nm 150 mg/L	*ci*B100	Kirloskar/MC-FIS; 1500 rpm, various loads	Ti-9	[53]
	20 nm 100 ppm	*ci*B100	Simpson-S217/MC-FIS; 1200 rpm, full load	Ti-10	[10]
	20 nm 100 ppm	*ci*B100	Simpson-S217/MC-FIS; 1200 rpm, full load	Ti-11	[80]
	60 nm 200 ppm	*m*B100	Kirloskar TV1/MC-FIS; 1800 rpm, full load	Ti-12	[87]
	50 nm 200 ppm	*mo*B100	Kirloskar/MC-FIS; 1500 rpm, full load	Ti-13	[140]
	<100 nm 0.01 wt%	D/*wco*B20/Bu10	3 LD 510/DI; 2800 rpm, various loads	Ti-14	[85]

NO_x emission, especially at higher loads, due to the higher in-cylinder temperature [21, 73].

Addition of Al_2O_3 NPs and AlOOH NPs to various biodiesels, emulsions, and blends resulted mainly in improved engine characteristics, Fig. 5.7. Anchupogu et al. [6] tested Al_2O_3 NPs added to the D/*ci*B20 base fuel. They reported that with a dosage of 40 ppm, the NO_x emission can be reduced due to Al_2O_3 NPs ability to scavenge nitric oxide radicals. Furthermore, by using Al_2O_3-D/*ci*B20 with or without EGR, the HC emission can be reduced due to catalytic effect of NPs, which lowers the carbon

Table 5.9 Test cases with Zn-based NMs

Nano-additive	NMs: size dosage	Base fuel	Diesel engine/FIS; operating conditions	Case name	References
ZnO NPs	67 nm 100 ppm	D100	Kirloskar AV1/MC-FIS; 1500 rpm, full load	Zn-1	[56]
	<100 nm 100 ppm	D100	Diesel engine/MC-FIS; 1500 rpm, full load	Zn-2	[118]
	34 nm 50 ppm	D/W10	Diesel engine/MC-FIS; 3000 rpm, full load	Zn-3	[43]
	<100 nm 100 ppm	D/*m*B20	Diesel engine/MC-FIS; 1500 rpm, full load	Zn-4	[118]
	<100 nm 100 ppm	D/*m*B50	Diesel engine/MC-FIS; 1500 rpm, full load	Zn-5	[118]
	35 nm 100 ppm	*ci*B100	Simpson-S217/MC-FIS; 1200 rpm, full load	Zn-6	[80]
	50 nm 100 ppm	*ci*B100	Simpson-S217/MC-FIS; 2000 rpm, full load	Zn-7	[8, 9]
	30 nm 250 ppm	D/*p*B30/E30	Kirloskar/MC-FIS; 1500 rpm, full load	Zn-8	[93]

activation temperature and enhances the combustion process. The smoke emission can also be reduced due to shorter ignition delay, caused by a higher cetane number and better fuel/air mixing of the tested nanofuel. El-Seesy et al. [35] reported that lower NO_x emission, obtained in case Al-12, is attributed to shortened combustion duration due to the catalytic effect of Al_2O_3 NPs. Similar results were obtained in case Al-23 by Nour et al. [82] when testing the Al_2O_3-D/*jj*B10/E20 nanofuel at high engine loads. Namely, the NO_x emission was observed to decrease strongly (compared to the base fuel) due to the catalytic effect of Al_2O_3 NPs; furthermore, the existence of lower active radicals reduced the chances for thermal NO_x formation. Meanwhile, at low engine loads no obvious trend regarding the NO_x emission was observed. In case Al-7, Al_2O_3 NPs can also act as NO_x reduction agents by converting NO_x into oxygen and N_2 during the combustion process [43]. In cases Al-13, Al-14, Al-16, Al-18, Al-19, and Al-22, the Al_2O_3 NPs addition increases NO_x emissions due to higher maximum heat release rate and higher maximum in-cylinder pressure, caused by the catalytic effect of NPs which accelerates the combustion process. Aalam and Saravanan [1] showed that the NO_x emission notably increases by increased dosage of Al_2O_3 NPs in the D/*m*B20 base fuel up to 100 ppm; this is because Al_2O_3 NPs act as an oxygen-donating catalyst. Figure 5.7 shows that in all cases HC, CO, smoke, and PM emissions, g_e, P_e, and η_{th} are improved. The biggest improvement of g_e and η_{th} was obtained in case Al-21, while the highest reduction of HC, CO, smoke, and PM emissions was reached in cases Al-17, Al-12, Al-17, and Al-7, respectively. By comparing all cases involving aluminum-based NPs, the most promising overall results were obtained by addition of Al_2O_3 NPs to D/*jj*B20 (case Al-21) and to D/E35 (case Al-12).

Table 5.10 Test cases with C NTs

Nano-additive	Nano-additive: size dosage	Base fuel	Diesel engine/FIS; operating conditions	Case name	References
C NTs	40–60 nm 100 ppm	D100	Yanmar TF120M/DI; 1800 rpm, full load	C-1	[21]
	<100 nm 100 ppm	D100	Diesel VCR/DI; 1500 rpm, full load	C-2	[132]
	<100 nm 50 ppm	D/*h*B20	Kirloskar/MC-FIS; 1500 rpm, 80% load	C-3	[113]
	<100 nm 100 ppm	D/*s*B50	Diesel VCR/DI; 1500 rpm, full load	C-4	[132]
	33 nm 60 ppm	D/*ts*B20	Kirloskar TV1/MC-FIS; 1500 rpm, full load	C-5	[100]
	4–8 nm 60 ppm	D/*wco*B2	DICOM 50/DI; various speeds, full load	C-6	[44]
	<100 nm 100 ppm	*s*B100	Diesel VCR/DI; 1500 rpm, full load	C-7	[132]
	<200 nm 100 ppm	*j*B/*tp*O10	Greaves Cotton 5520/MC-FIS; 3000 rpm, full load	C-8	[121]
	16 nm 100 ppm	*j*B/W5	Kirloskar TAF1/MC-FIS; 1500 rpm, full load	C-9	[15]
	5–20 nm 100 ppm	*wco*B100	NA/MC-FIS; 1000 rpm, full load	C-10	[79]

5.5.3 NMs Based on Cerium

The cases involving CeO_2 NPs are collected in Table 5.3. The corresponding reported variations of engine characteristics are shown in Fig. 5.8. Vairamuthu et al. [131] studied the influence of:

- the number of nozzle holes and their diameters (3 × 0.28 mm, 4 × 0.22 mm, and 5 × 0.24 mm) and
- various dosages (ranging from 20 to 60 ppm) of CeO_2 NPs added to the D/*ci*B50 (case Ce-4) and *ci*B100 base fuels (case Ce-12).

They observed that the best engine characteristics were obtained by using 3 nozzle holes with a diameter of 0.28 mm with 40 ppm of CeO_2 NPs in the *ci*B100 base fuel. In case Ce-5, Patil et al. [91] investigated the influence of various compression rates of a diesel engine and various contents of CeO_2 NPs in the D/csB10 base fuel. They showed that the best engine characteristics are achieved by adding 50 ppm of CeO_2 NPs to D/csB10 at a compression ratio of 14.

Table 5.11 Test cases with carbon-, graphite-, and graphene-based NMs

Nano-additive	NMs: size dosage	Base fuel	Diesel engine/FIS; operating conditions	Case name	References
swC NTs	1–2 nm 25 ppm	D100	HATZ 1B30/DI; 2000 rpm, full load	wC-1	[84]
mwC NTs	10–15 nm 30 mg/L	D/*jj*B20	HATZ-IB30-2/DI; 2000 rpm, full load	wC-2	[32]
	10–20 nm 90 ppm	D/*wco*B10	Lombardini 3LD510/CR-FIS; 2800 rpm, full load	wC-3	[48]
	5 nm 30 ppm	D/*wco*B20	Diesel engine/DI; 1000 rpm, full load	wC-4	[39]
	10–15 nm 50 mg/L	*j*B/Bu40	HATZ-1B30-2/DI; 2000 rpm, full load	wC-5	[31]
G'O NPs	1–2 nm 25 ppm	D100	HATZ 1B30/DI; 2000 rpm, full load	G'-1	[83]
Cqd NPs	1–3 nm 60 ppm	D/*wco*B15/W1	Lombardini 3LD 510/DI; 2700 rpm/full load	Cq-1	[36]
G NPTs	6–8 nm 100 mg/L	D/*j*B20	HATZ-1B30-2/DI; 2000 rpm, full load	G-1	[33]
	22–26 nm 40 ppm	D/*sg*B20	Kirloskar TV1/MC-FIS; 1500 rpm, full load	G-2	[89]
	6–8 nm 50 mg/L	*j*B/Bu40	HATZ-1B30-2/DI; 2000 rpm, full load	G-3	[31]
GO NPs	6 nm 50 mg/L	*j*B100	HATZ-1B30-2/DI; 2000 rpm, full load	G-4	[34]
	1–2 nm 90 ppm	D/*aa*B20	Lombardini 3LD 510/DI; 2100 rpm, various loads	G-5	[46]
	10–15 nm 50 mg/L	*j*B/Bu40	HATZ-1B30-2/DI; 2000 rpm, full load	G-6	[31]

Addition of CeO_2 NPs to various base fuels results mainly in improved engine characteristics. Only Ooi et al. [84] reported that in case Ce-2, the NO_x emission is increased due to higher energy content and higher combustion temperature. In cases where the CeO_2 NPs are added in the form of a nanofluid (cases Ce-6 and Ce-15) or the base fuel contains water (cases Ce-3, Ce-9, Ce-10, Ce-11, Ce-15), NO_x emission is reduced due to increased nanofuel heat absorption capacity and consequent reduction of the in-cylinder temperature. In other cases, the CeO_2 NPs acted as NO_x reduction agents by converting NO_x into oxygen and N_2 during the combustion process. The reactivity role of CeO_2 NPs, acting as oxidation agents due to their oxygen content, secondary atomization by micro-explosions, and catalytic activity were reported as the most important parameters for the reduction of CO, HC,

Table 5.12 Test cases with organic NMs

Nano-additive	NMs: size dosage	Base fuel	Diesel engine/FIS; operating conditions	Case name	References
CS NPs	20 nm 0.1 wt%	D/*pp*B20	Kirloskar TV1/MC-FIS: 1500 rpm, full load	CS-1	[136]
RH NPs	<100 nm 0.1 wt%	D/*po*B20	Kirloskar TV1/MC-FIS: 1500 rpm, full load	RH-1	[88]
	20 nm 0.1 wt%	D/*pp*B20	Kirloskar TV1/MC-FIS: 1500 rpm, full load	RH-2	[137]

smoke, and PM emissions and specific fuel consumption as well as for the increase of engine power and thermal efficiency. By comparing all cases involving CeO_2 NPs, the most promising overall performance was obtained by addition of CeO_2 NPs to D/*wco*B5/W3 (case Ce-10).

5.5.4 NMs Based on Cobalt

The cases involving Co_3O_4 NPs are collected in Table 5.4. The corresponding reported variations of engine characteristics are shown in Fig. 5.9.

By adding Co_3O_4 NPs to neat *ci*B100 biodiesel, the NO_x emission at full load is higher than without NPs. Meanwhile, the NO_x reduction in case Co-4 is attributed to reduced in-cylinder temperature due to the heat sink effect. Furthermore, the NO_x reduction when adding Co_3O_4 NPs to the blends of mineral diesel and biodiesel (cases Co-1 and Co-2) can be explained by the higher cetane number, which causes less fuel to be burned during the premixed combustion phase; consequently, the maximum combustion in-cylinder temperature is decreased. Other engine characteristics are improved in all tested cases. The highest reduction of HC emission, up to 60%, is obtained by addition of Co_3O_4 NPs to *j*B100. This is achieved by the disruption of primary droplets containing NPs, which results in secondary atomization, better mixing, and more complete combustion [108]. By comparing all cases involving Co_3O_4 NPs, the overall worst engine characteristics were obtained in the case Co-3.

Table 5.13 Test cases with other NMs

Nano-additive	NMs: size dosage	Base fuel	Diesel engine/FIS; operating conditions	Case name	References
B_2O_3 NPs	80–100 nm 0.5 wt%	D100	Kirloskar TV1/MC-FIS; 1500 rpm, full load	B-1	[73]
MnO NPs	33 nm 50 ppm	D/W10	Diesel engine/MC-FIS; 3000 rpm, full load	Mn-1	[43]
Mn_2O_3 NPs	30 nm 50 ppm	D/*wfo*B20	Diesel engine, urea SCR/DI; 1500 rpm, full load	Mn-2	[72]
NiO NPs	<20 nm 40 ppm	D/*p*B20	Comet VCR/MC-FIS; 1500 rpm, various load	Ni-1	[127]
SiO_2 NPs	20–30 nm 100 ppm	D100	Yanmar TF120M/DI; 1800 rpm, full load	Si-1	[21]
	10–30 nm 50 ppm	*r*B100	Mitsubishi Canter 4D34-2A/DI; various speeds, full load	Si-2	[86]
Al_2O_3 and CeO_2 NPs	51 and 32 nm 30 ppm	D/*j*B20	Kirloskar/DI; 1500 rpm, full load	AlCe-1	[94, 95]
	51 and 32 nm 30 ppm	*j*B100	Kirloskar/DI; 1500 rpm, full load	AlCe-2	[94, 95]
SiO_2 and TiO_2 NPs	50 nm 100 ppm	D/*bba*B20	NA engine/DI; 1500 rpm, full load	SiTi-1	[59]
Al@C NC	60–160 nm 30 ppm	D/*p*B10/E4	Cummins ISBe5/CR-FIS; ESC test	AlC-1	[139]
	50–60 nm 30 ppm	D/E4	Cummins ISBe5/CR-FIS; ECS test	AlC-2	[54]
CeO_2/mwC (DMF) NTs	40–50 nm 90 ppm	D/*wco*B20	Heavy duty OM355/MC-FIS; 13 mode tests	CeC-1	[75]

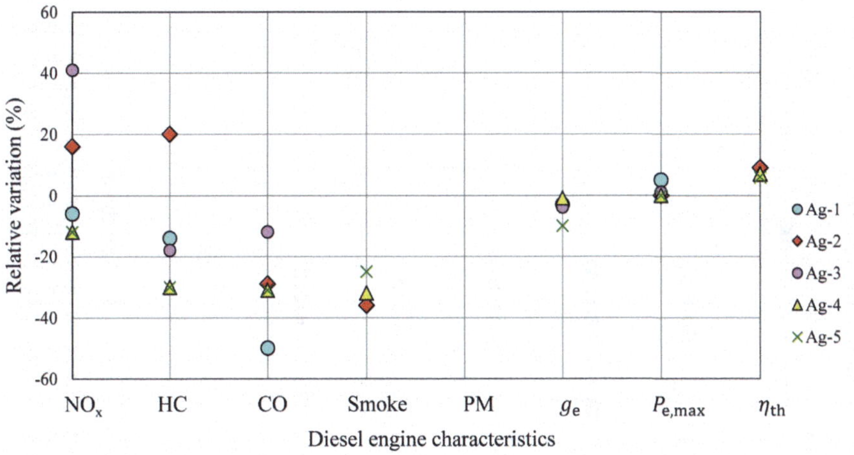

Fig. 5.5 Variations of engine characteristics due to addition of Ag-based NMs

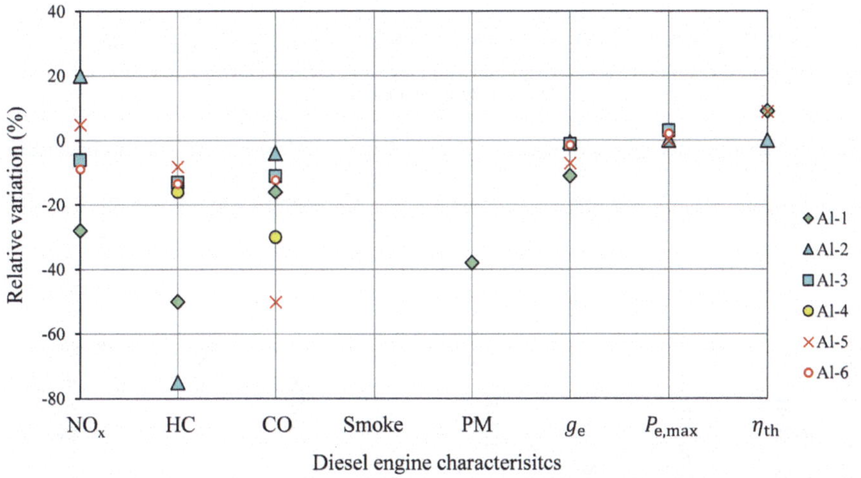

Fig. 5.6 Variations of engine characteristics due to addition of Al-based NMs (D100 base fuel)

5.5.5 NMs Based on Copper

The cases involving CuO NPs are collected in Table 5.5. The corresponding reported variations of engine characteristics are shown in Fig. 5.10. D'Silva et al. [23] investigated the addition of CuO NPs at dosages of 10, 30, and 50 mg/L added to D/*pp*B20 base fuel in a VCR diesel engine with variable compression ratio ranging from 12 to 18. The best engine characteristics were obtained at a fixed compression ratio of 17.5 with a dosage of 50 mg/L.

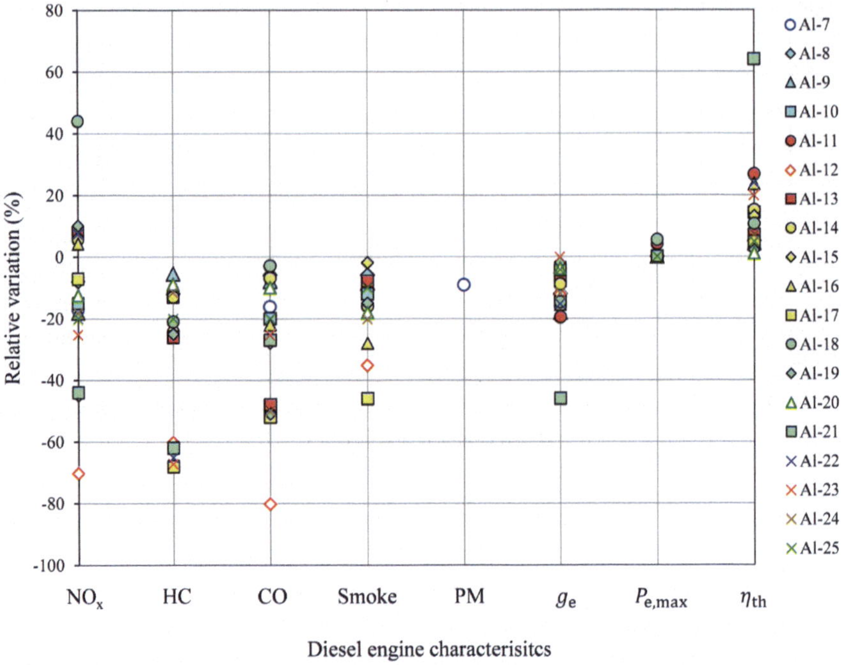

Fig. 5.7 Variations of engine characteristics due to addition of Al-based NMs (various base fuels)

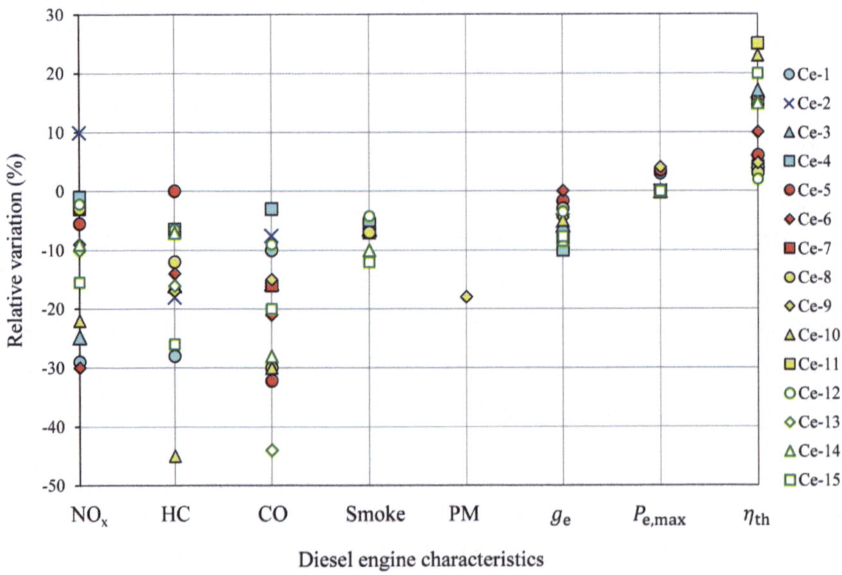

Fig. 5.8 Variations of engine characteristics due to addition of Ce-based NMs

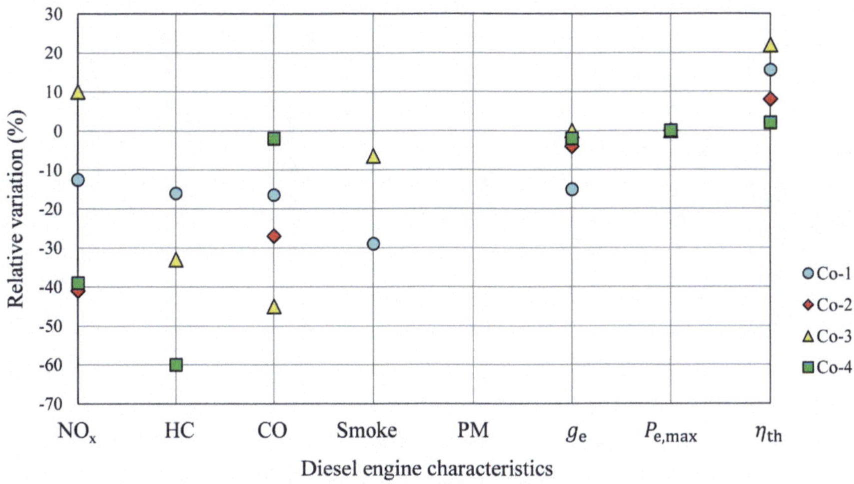

Fig. 5.9 Variations of engine characteristics due to addition of Co-based NMs

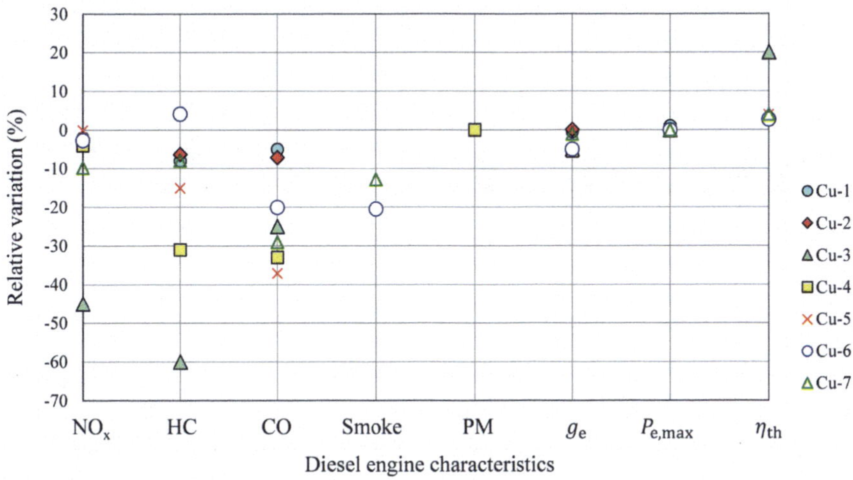

Fig. 5.10 Variations of engine characteristics due to addition of Cu-based NMs

Practically, all diesel engine characteristics were improved when CuO NPs were added to various base fuels. An exception was the HC emission which was slightly higher in the case Cu-6. A comparison of all cases shows that the maximum improvements of engine characteristics were reached in case Cu-3, where CuO NPs were added to D100. This is mostly attributed to the heat sink effect of the CuO NPs and their ability to act as NO_x reduction agents by converting NO_x into oxygen and N_2 during the combustion process. The catalytic effect of CuO NPs was the dominant one in reduction of HC and CO emissions in the case Cu-3.

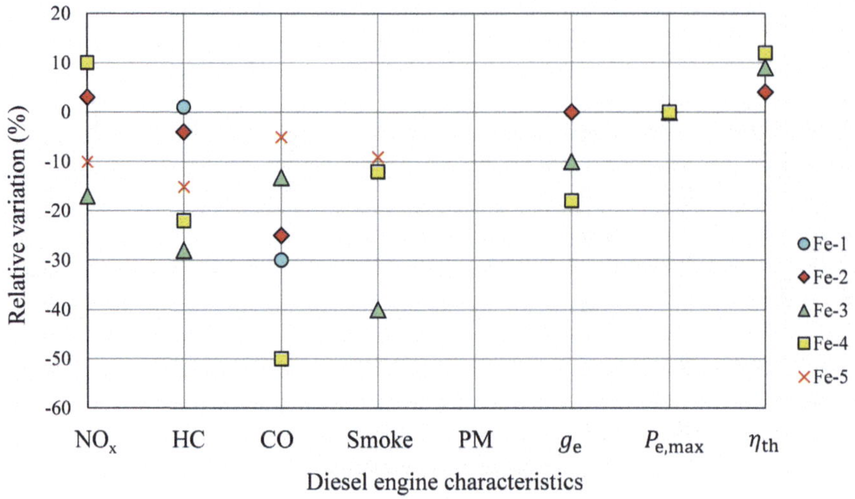

Fig. 5.11 Variations of engine characteristics due to addition of Fe-based NMs

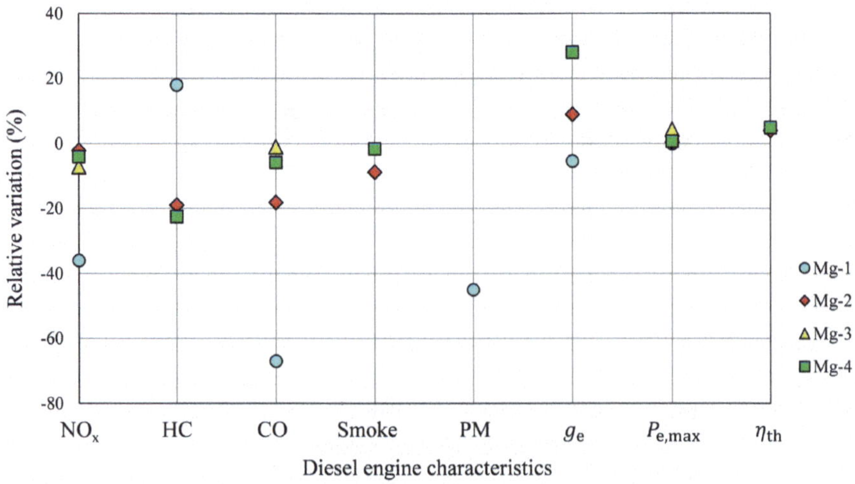

Fig. 5.12 Variations of engine characteristics due to addition of Mg-based NMs

5.5.6 NMs Based on Iron

Among iron-based NMs, the most suitable fuel additives are maghemite and magnetite NPs. The cases involving those NPs are collected in Table 5.6. The corresponding reported variations of engine characteristics are shown in Fig. 5.11.

Kumar et al. [66] investigated the addition of γ-Fe_2O_3 NPs, in the form of a nanofluid, to the D/*pp*B20 base fuel. They explained the reduction of NO_x emission

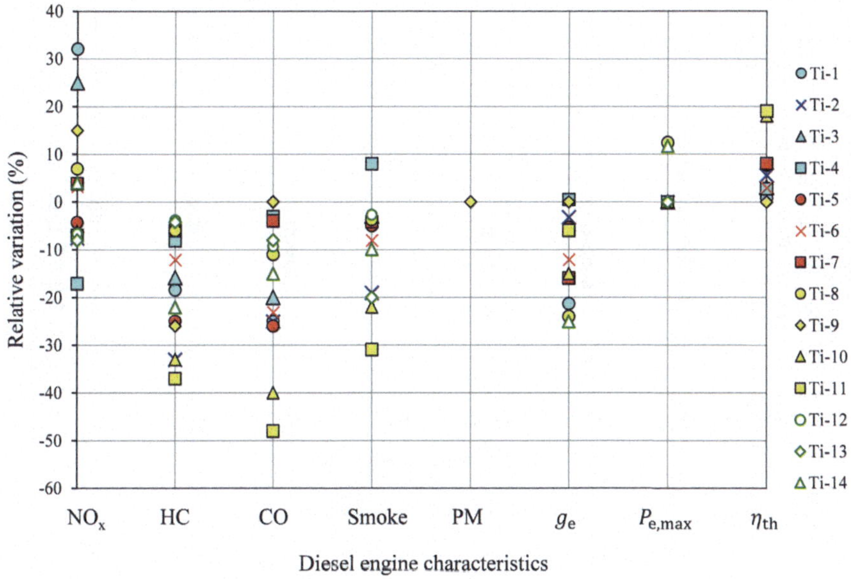

Fig. 5.13 Variations of engine characteristics due to addition of Ti-based NMs

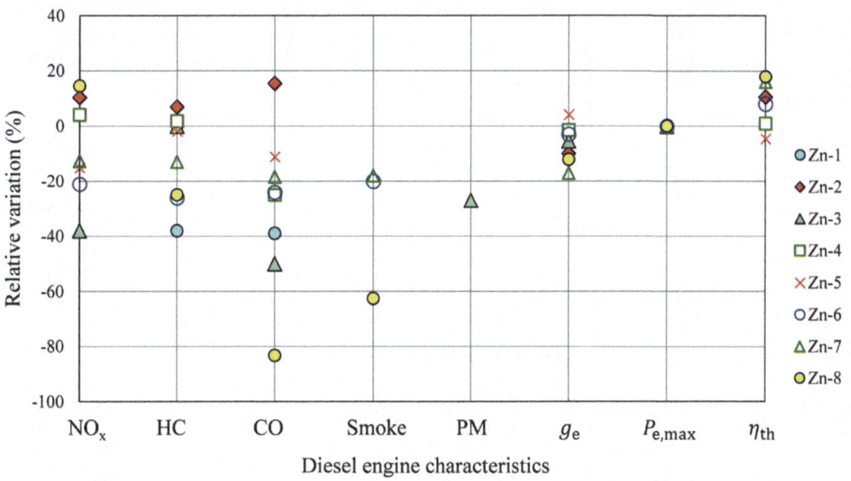

Fig. 5.14 Variations of engine characteristics due to addition of Zn-based NMs

by the increased convective heat transfer inside the combustion chamber, which reduces the average in-cylinder temperature. Furthermore, high latent heat and high thermal capacity of water in the nanofuel also helped in reducing the average in-cylinder temperature and consequently also NO_x emission in the case Fe-3 [66]. The highest reduction of HC emission is also achieved in the Fe-3 case, where the

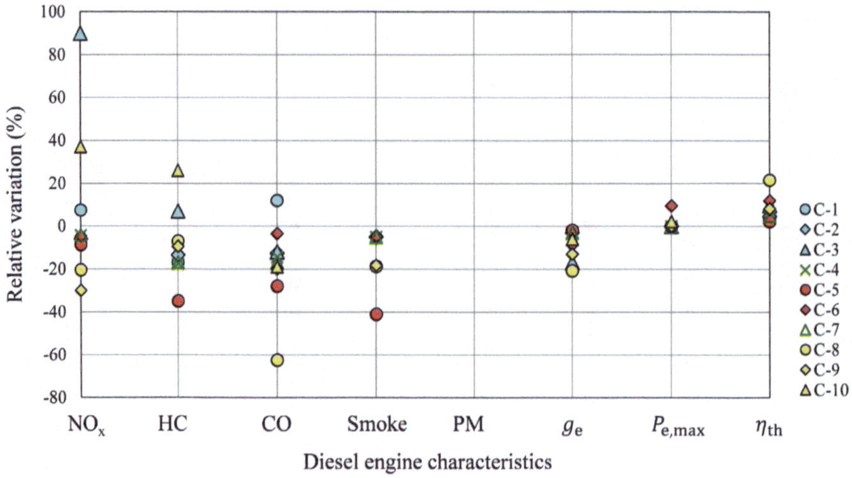

Fig. 5.15 Variations of engine characteristics due to addition of C NTs

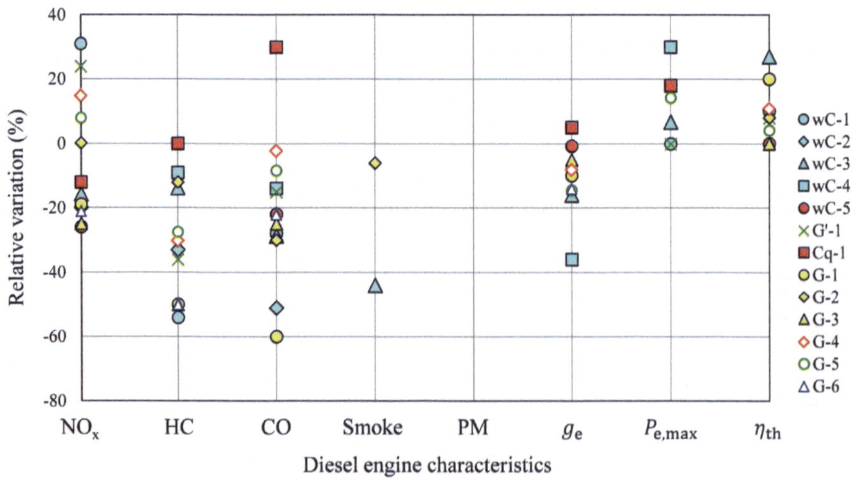

Fig. 5.16 Variations of engine characteristics due to addition of carbon-, graphite-, and graphene-based NMs

reactivity role of γ-Fe_2O_3 NPs, acting as oxidation agents due to their oxygen content, is highlighted. The highest reduction of CO emission was observed in the case Fe-4, where the CO emission decreases with the increase of the burnt fuel fraction in the premixed combustion phase, due to chemical reactivity, and catalytic effects of γ-Fe_2O_3 NPs [77]. On the other hand, case Fe-4 also exhibits the highest increase of NO_x emission due to higher in-cylinder temperature and shortened ignition delay. By comparing all cases involving γ-Fe_2O_3 NPs, the lowest NO_x, HC, and smoke emissions, accompanied by acceptable other engine characteristics, are obtained in

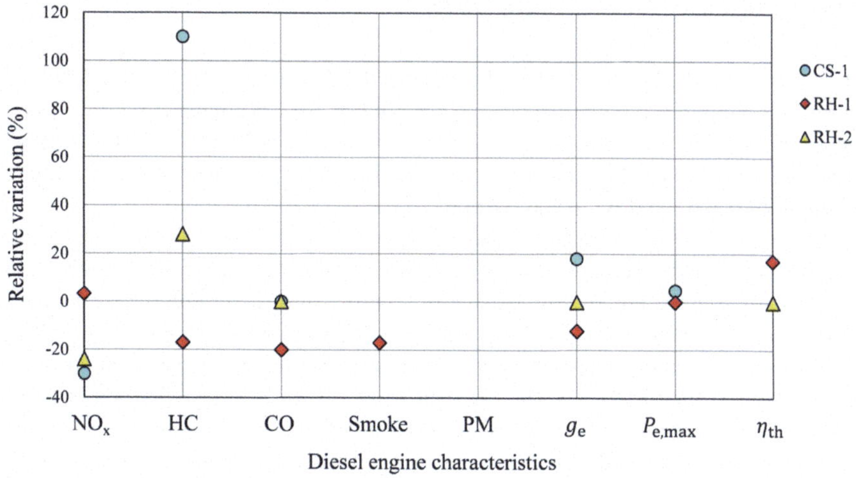

Fig. 5.17 Variations of engine characteristics due to addition of organic NMs

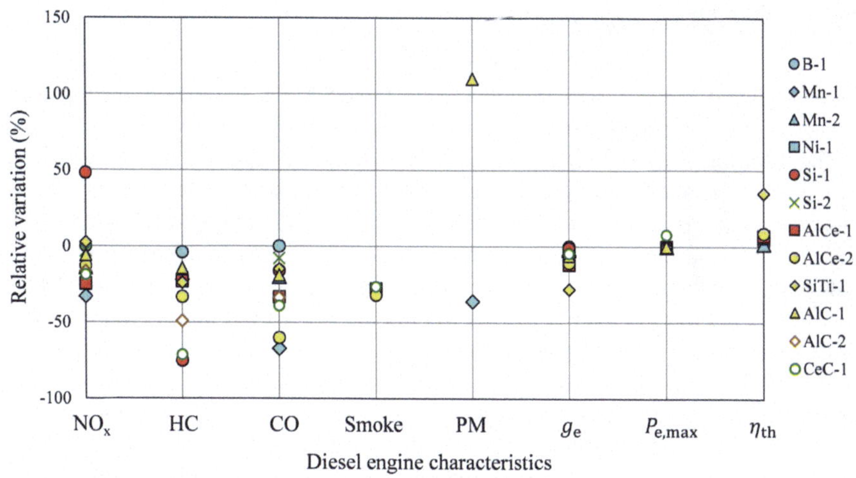

Fig. 5.18 Variations of engine characteristics due to addition of other NMs

the case Fe-3. It follows that the addition of 30-nm-sized γ-Fe_2O_3 NPs in a dosage of 1 wt% to D/*pp*B20 base fuel can be viewed as the best iron-based nanofuel for the tested engine [77].

5.5.7 NMs Based on Magnesium

The cases involving MgO NPs are collected in Table 5.7. The corresponding reported variations of engine characteristics are shown in Fig. 5.12. Ranjan et al. [106] tested the addition of MgO to D/*wco*B20 (case Mg-2) and to *wco*B100 (case Mg-4) in the same engine at the same operating conditions. The particles were sized in the range of 20–30 nm, and the dosage was 30 ppm.

Ranjan et al. [106] showed that the addition of MgO NPs to the D/*wco*B20 and *wco*B100 base fuels increases g_e due to higher densities of nanofuels, which leads to higher mass injection (at the same fuel volume) into the combustion chamber. Furthermore, Ranjan et al. [106] reported increased P_e, which in cases Mg-2 and Mg-4 was attributed to the reduced pump leakage due to higher viscosity and density of the engaged nanofuels. As can be seen from Fig. 5.12, the reductions of CO and smoke emissions in the case Mg-2 are higher than those in the case Mg-4; this was attributed to higher oxygen content and better secondary atomization of MgO-D/*wco*B20 when compared to MgO-*wco*B100.

5.5.8 NMs Based on Titanium

The cases involving TiO_2 NPs are collected in Table 5.8. The corresponding reported variations of engine characteristics are shown in Fig. 5.13.

The addition of TiO_2 NPs to various base fuels results mainly in improved engine characteristics with the exception of the NO_x emission. TiO_2 NPs have been found capable of enabling hydroxyl radical reactions with unburned HC and further oxidizing HC into H_2O and CO_2. Lawrence et al. [70] investigated the effect of 20% EGR and SCR technologies on engine characteristics at full load by using TiO_2-D/*me*B20 nanofuel. They reported that without engaging EGR and SCR, TiO_2 NPs addition increases the NO_x up to 25% and decreases HC, CO, and smoke emissions up to 16%, 20%, and 3%, respectively. Furthermore, g_e is slightly decreased and η_{th} is increased by 3%. The added EGR decreases NO_x, HC, and CO emissions as well as η_{th}, while smoke emission and g_e increase. A further reduction of all emissions was observed by adding the SCR technology. As reported, by running the engine with TiO_2-D/*me*B20 nanofuel with 20% EGR and SCR with 35% urea concentration, the NO_x, HC, CO, and smoke emissions decreased up to 60%, 35%, 54%, and more than 50%, respectively [70]. Praveen et al. [97] investigated the addition of TiO_2 NPs to D/*ci*B20 on a diesel engine without (case Ti-6) and with EGR (case Ti-7). From Fig. 5.13, it is evident that TiO_2 NPs addition has more or less similar effects on engine characteristics, and the results seem to be rather independent of EGR engagement. After investigating TiO_2-D/*wco*B20 and TiO_2-D/*wco*B20/Bu10, Örs et al. [85] reported that the catalytic effects of TiO_2 NPs are responsible for increased NO_x emission due to increased in-cylinder pressure and temperature. In the case Ti-14, where TiO_2 NPs were added to D/*wco*B20/Bu10, the reduction of harmful emissions is higher than in case Ti-8. This was attributed to a combination of butanol and TiO_2 NPs which enhances micro-explosions and consequently secondary atomization. Among all considered cases, the addition of TiO_2 NPs to *Calophyllum inophyllum* biodiesel (cases Ti-10 and Ti-11) exhibits the most favorable compromise between harmful emissions, specific fuel consumption, engine power, and thermal efficiency.

5.5.9 NMs Based on Zinc

The cases involving ZnO NPs are collected in Table 5.9. The corresponding reported variations of engine characteristics are shown in Fig. 5.14.

In general, if the base fuel contains water (case Zn-3) or alcohol (case Zn-8), or the NMs additive is a nanofluid (cases Zn-6 and Zn-7), the heat absorption induced by water/alcohol molecules and the resulting heat sink effect leads to a reduction of NO_x emission. The opposite, however, was observed in case Zn-8 since the heat sink effect was outweighed by other effects: micro-explosions, oxidizing reagent activity, and catalytic activity of ZnO NPs as well as higher calorific value and cetane number of the ZnO-D/*p*B30/E30 nanofuel. In the case Zn-7, where ZnO NPs were added to the *ci*B100 base fuel, the reduced smoke emission was attributed to the NPs catalytic activity which was found to enhance chemical reactions during the combustion process [8, 9]. By comparing all cases involving ZnO NPs, the most promising engine characteristics were obtained by addition of ZnO NPs to D/W10 (case Zn-3) and to *ci*B100 (case Zn-7).

5.5.10 Carbon NTs

The cases involving C NTs added to various base fuels are collected in Table 5.10. The corresponding reported variations of engine characteristics are shown in Fig. 5.15.

The maximal NO_x emission reduction was obtained in the case C-9; Basha and Anand [15] explained this by the fact that encapsulated C NTs, added in the form of a nanofluid, into the *j*B/W5 base fuel can absorb heat very rapidly. This decreases the burning gas temperature inside the combustion chamber and results in lower NO_x emission. On the contrary, the maximum NO_x emission increase was observed by Sandeep et al. [113] in the case C-3 where NO_x emissions were found to increase drastically by increasing engine loads due to shortened ignition delay, higher premixed combustion fraction, and higher peak in-cylinder temperature. The maximal reductions of HC and smoke emissions were obtained in case C-5 due to the reactivity and catalytic effect of C NTs and consequent enhanced combustion [100]. In the case C-9, where the C-*j*B/W5 nanofuel was tested at various engine loads, secondary atomization by micro-explosions was reported as the main reason for reduced HC emission [15]. The maximal reduction of CO emission was reported by Sharma et al.

[121] due to C NTs acting as a catalyst to accelerate the burning rate. This results in decreased ignition delay, lower heat release rate, and advanced peak of the heat release rate. Maximal thermal efficiency and minimal specific fuel consumption were reported by Sharma et al. [121] in the case C-8 and attributed to better atomization caused by addition of C NTs to *j*B/*tp*O10 base fuel. By comparing all cases involving C NTs, the most promising engine characteristics were obtained by addition of C NTs to D/*ts*B20 (case C-5) and to *j*B/*tp*O10 (case C-8).

5.5.11 NMs Based on Carbon, Graphite, and Graphene

The cases involving swC and mwC NTs, G'O NPs, Cqd NPs, G NTPs, and GO NPs addition to various base fuels are collected in Table 5.11. The corresponding reported variations of engine characteristics are shown in Fig. 5.16.

The highest NO_x emission was reported by Ooi et al. [84] in case wC-1, where the addition of the swC NTs to D100 base fuel increases the in-cylinder temperature due to higher energy content of swC-D100 nanofuel. The lowest NO_x emission was achieved in case wC-5, where mwC NTs were added to *j*B/Bu40 nanofuel [31]. In the case wC-1, the minimal HC emission was observed due to advanced start of combustion and shorter combustion duration [84]. The maximal g_e and CO emissions were observed in case Cq-1, where Cqg NPs were added to D/*wco*B15/W1 base fuel [36]. The lowest CO emission was reached in the case G-1, where G NPTs were added to D/*j*B20 base fuel. The main reason for this is the strong catalytic effect of G NPTs due to their high surface area-to-volume ratio, which improves the fuel/air mixing in the combustion process [33]. The results of smoke emission were reported only in the cases wC-3 and G-2. The lower smoke emission in case wC-3, where mwC NTs were added to D/*wco*B10 base fuel, was explained by enhanced chemical reactions during the combustion process due to the NTs catalytic activity [48]. By comparing all cases involving swC and mwC NTs, graphite- and graphene-based NMs, the overall most promising engine characteristics were obtained by addition of mwC NTs to D/*wco*B10 (case wC-3).

5.5.12 Organic NMs

The cases involving organic NPs (coconut shell and rice husk) are collected in Table 5.12. The corresponding reported variations of engine characteristics are shown in Fig. 5.17. All considered cases were tested on the same diesel engine and FIS and at similar operating conditions.

In case CS-1, the addition of CS NPs to D/*pp*B20 increased significantly the HC emission at full load while at idling condition and intermediate load the HC emission was reduced [136]. The addition of RH NPs (case RH-1) helps in reducing the smoke, HC, and CO emissions, since the CS NPs improve the combustion process by providing excess oxygen to the base fuel [88]. The improved combustion process leads to lower g_e and higher η_{th}. Furthermore, better oxidation also increases the in-cylinder temperature, thereby increasing slightly also the NO_x emission [88]. By comparing organic NMs fuel additives, the best engine characteristics were observed in the case RH-1, where RH NPs were added to D/*po*B20 base fuel.

5.5.13 Other NMs

The cases involving rather rarely tested NMs or NMs combinations are collected in Table 5.13. These cases involve: boron oxide (case B-1), manganese oxides (cases Mn-1 and Mn-2), nickel oxide (case Ni-1), silicon dioxide (cases Si-1 and Si-2), mixture of Al_2O_3 and CeO_2 NPs (cases AlCe-1 and AlCe-2) and of SiO_2 and TiO_2 NPs (case SiTi-1), Al@C core@shell nanocomposite (cases AlC-1 and AlC-2), and functionalized CeO_2/mwC (DMF) NTs. The corresponding reported variations of engine characteristics are shown in Fig. 5.18.

Wu et al. [139] reported that the PM count emission increases significantly by addition of Al@C NPs to the D/pB19/E4 base fuel. They assumed that Al@C NPs burn during the combustion process in order to form alumina NPs which results in substantial increase of emitted PM count. By comparing all considered cases, the most promising engine characteristics were obtained by addition of MnO NPs to D/W10 base fuel (case Mn-1). Namely, MnO NPs have a potential to reduce PM emission, especially at higher loads, due to higher evaporation rate and shorter ignition delay; shorter ignition delay may reduce the premixed burn duration [43].

5.5.14 Overview of Nanofuel Influence on Diesel Engine Characteristics

The effects of NMs fuel additives on the considered diesel engine characteristics can be summarized as follows.

NO_x emission. As can be seen from Tables 5.1, 5.2, 5.3, 5.4, 5.5, 5.6, 5.7, 5.8, 5.9, 5.10, 5.11, 5.12, and 5.13, NMs addition can influence NO_x emission either way, positively or negatively. Roughly speaking, about one-third of various nanofuel investigations showed an increase in NO_x emission, while the other two-thirds indicate more or less positive NO_x emission effects. As reported, the NO_x emission may increase due to higher heat release rate and higher in-cylinder pressure and temperature. As reported, this may be caused by higher calorific value and higher oxygen content of the nanofuel [1, 93, 100] and by the NMs catalytic effects which accelerate the combustion process and increase diffusion-controlled combustion [1, 2, 34, 85, 125]. On the other hand, the main reason for NO_x emission reduction is attributed to NMs heat sink effect, especially when NMs are added to the base fuel in the form of a nanofluid [13, 15, 19, 26–28, 60, 61, 64, 66, 78, 99, 101, 107–109, 114, 140]. When the additive is a nanofluid or the base fuels contain water or alcohol, the heat absorption by water/alcohol molecules can lead to micro-explosions; these may also help in reducing the NO_x emission [8]. Furthermore, some NMs can also act as NO_x reduction agents by converting NO_x into oxygen and N_2 during the combustion process [43, 68, 80, 94, 98, 103, 106, 111, 118, 131].

HC emission. NMs addition to various nanofuels mainly decreases the HC emission, Tables 5.1, 5.2, 5.3, 5.4, 5.5, 5.6, 5.7, 5.8, 5.9, 5.10, 5.11, 5.12, and 5.13. HC emission can decrease due to the NMs reactivity role, where the NMs act as oxidation agents due to their oxygen content [1, 2, 8, 9, 26, 29, 43, 46, 66, 70, 80, 85, 87, 92, 94, 101, 111, 115, 118, 119, 125, 131, 140] and due to accelerated combustion, caused by the NMs catalytic effect [6, 31–35, 59–61, 78, 79, 98, 106, 113, 115]. Furthermore, in the case of water-in-diesel emulsion base fuels, the dominant effect for HC reduction is secondary atomization by micro-explosions [13, 15, 64, 107]. When carbon is present in the NMs structure, NMs can also increase the HC emission. This was confirmed in [79], where it was reported that the HC emission, obtained by using the C-*wco*B100 nanofuel, increased with respect to those of the *wco*B100 base fuel.

CO emission. CO emission reduction has been frequently and more typically reported when various NMs are added to various base fuels, although in some cases CO emission increase was also observed, Tables 5.1, 5.2, 5.3, 5.4, 5.5, 5.6, 5.7, 5.8, 5.9, 5.10, 5.11, 5.12, and 5.13. CO emission can decrease due to NMs catalytic role [1, 2, 6, 13, 15, 29, 32–35, 45, 59, 61, 78, 79, 125, 132, 140], reactivity role, where the NMs act as oxidation agents which convert CO into CO_2 during the combustion process [9, 27, 28, 43, 46, 55, 64, 66, 67, 70, 80, 82, 85, 87, 91, 94, 95, 98–101, 103, 106, 108, 119, 121, 122, 125, 131], and secondary atomization due to micro-explosions [55, 107]. In a few investigations, a negative effect of NMs fuel additives on CO emission was reported. The main reasons for increased CO emission are additional carbon present in NMs and the presence of oxygenic functional groups [36].

PM emission. Unfortunately, practically all NMs engaged as diesel engine fuel additives end up as particulates (in one or another form) in the diesel engine exhaust. Consequently, the increase of the number of particulates (PM count) is typically substantial, whereas the contribution to the total mass of the particulates (PM mass) is usually not significant due to extremely low mass of NMs particles. Under special circumstances, nanofuels may even deliver lower total PM mass due to the NMs activities, for example, by promoting soot oxidation. The total PM mass can be reduced by addition of NMs, which provide additional oxygen that enables reactions with C atoms, and promote secondary atomization, caused by micro-explosions and due to the catalytic activity of NMs [19, 55]. It should be noted that NMs are by themselves a direct source of particulates, but besides this, addition of metallic NMs into the base fuel may also result in formation of new self-nucleated metallic particles [139]. Furthermore, Zhang and Balasubramanian [142] found that the usage of nanofuels, enriched with Ce- and Fe-based NMs, results in the formation of new particles within the peak diameter range of 10–20 nm (nucleation mode), leading to an increase in the total PM count emission.

Smoke emission. Smoke emission is actually a part of the total PM emission, but it is typically investigated and measured separately for various reasons. Anyhow, the experimentally obtained data is surely a valuable contribution to the NMs engagement evaluation. Practically, all tested nanofuels enable the reduction of smoke emission due to the shorter ignition delay and better fuel/air mixing [6, 68, 94, 95]. Furthermore, the reduced smoke emission was attributed to the NPs catalytic activity which was found to enhance chemical reactions during the combustion process [8, 9] and the oxygen present in the NMs [2, 88]. The importance of the oxygen contained within the NMs and the NMs reactivity with respect to smoke emission was confirmed by tests conducted with various nanofuels [35, 70, 80, 85, 87, 88, 99, 101, 105, 111, 121, 122, 125, 131]. For example, CeO_2 NPs can react directly with C atoms in soot, thereby lowering the oxidation temperature [111]. By adding various NPs (even carbon-based NMs) in the form of a nanofluid to the various base fuels, smoke emission typically reduces due to the enhanced secondary atomization effects, caused by micro-explosions [13, 15, 29, 107, 140].

Brake-specific fuel consumption. As this is the case with brake thermal efficiency, brake-specific fuel consumption (g_e) is also typically improved when NMs are used, Tables 5.1, 5.2, 5.3, 5.4, 5.5, 5.6, 5.7, 5.8, 5.9, 5.10, 5.11, 5.12, and 5.13. The most frequently reported reasons for the decreased g_e are the NMs catalytic effect [6, 8, 26–29, 34, 35, 92–95, 100], high thermal conductivity of nanofuels [132], high CN of nanofuels [46], and secondary atomization due to micro-explosions [13, 15, 36] [43, 94, 95]. Furthermore, a comparison of Al_2O_3 and CuO NPs, added to D100 base fuel in Lombardini 15LD 300 diesel engine [41], showed that Al_2O_3 NPs result in quite significant g_e decrease because they cause a stronger ignition delay reduction, which leads to more complete combustion. Only a few researchers reported that NMs additives increased g_e. For example, Ranjan et al. [106] showed that the addition of MgO NPs to the D/*wco*B20 and *wco*B100 base fuels increases g_e due to higher densities of nanofuels, which leads to higher mass injection (at the same fuel volume) into the combustion chamber.

Engine power. Typically, engine power (P_e) is not influenced significantly when various NMs are added to various base fuels, Tables 5.1, 5.2, 5.3, 5.4, 5.5, 5.6, 5.7, 5.8, 5.9, 5.10, 5.11, 5.12, and 5.13. Nevertheless, a few researchers reported about minor P_e increase under special circumstances. Chaichan et al. [19] reported a slight increase of P_e at various engine speeds at constant medium load. The NMs used were Al_2O_3 NPs, in the form of a nanofluid, added to the D100 base fuel, and P_e was reported to increase by increasing the NPs dosage. The reason stated was improved combustions because of micro-explosions, which enhance fuel/air mixing. Besides micro-explosions, Ettefaghi et al. in [36] quote high contacting surface and presence of oxygenic functional groups over the surface of Cqd NPs as the reason for P_e increase when using the Cqd-D/*wco*15/W1 nanofuel. Ranjan et al. [106] investigated the influence of 30 ppm of MgO NPs, added to the D/*wco*B20 and *wco*B100 base fuels. They reported increased P_e which was attributed to the reduced pump leakage due to higher viscosity and density of the engaged nanofuels.

Brake thermal efficiency. Practically, all investigations report that NMs fuel additives typically improve the brake thermal efficiency (η_{th}), Tables 5.1, 5.2, 5.3, 5.4, 5.5, 5.6, 5.7, 5.8, 5.9, 5.10, 5.11, 5.12, and 5.13. Only Seela et al. [118] reported that the ZnO NPs, added to D/*m*B50 base fuel, reduced η_{th} by about 4.5% at various engine loads. The typically listed reasons for improved η_{th} are higher calorific value and higher CN of the tested nanofuels (compared to the base fuels) [100], the catalytic activity of NMs [2, 7–9, 26–29, 33, 34, 59–61, 80–82, 88, 92–95, 97, 106, 119, 131], the oxygen content of NMs [45, 82, 127], and secondary atomization due to micro-explosions [3, 8, 9, 13, 15, 80]. For example, by comparing the effects of Al_2O_3 NPs and C NTs addition to the D/*ts*B20 base fuel, Raju et al. [100] reported that the oxygen present in the Al_2O_3 NPs takes part in the combustion reactions and thereby increases the combustion efficiency, which in turn results in higher η_{th}. Furthermore, the NMs, added in the form of a nanofluid, also resulted in higher η_{th} [98]; this was attributed to water presence in the nanofluid, which helps in fast evaporation and better fuel/air mixing.

Under the bottom line, one can say that the laboratory experiments published so far clearly indicate that nanofuels may offer a whole range of important benefits. As this is usually the case, these benefits are accompanied with additional cost and unwanted side effects. In spite of that, one can be sure that NMs fuel additives are worth of investigating their potential benefits and practical viability of long-term usage in diesel engines.

5.6 Identification of Best Performing Nanofuels

A glance at Tables 5.1, 5.2, 5.3, 5.4, 5.5, 5.6, 5.7, 5.8, 5.9, 5.10, 5.11, 5.12, and 5.13 reveals that in the search for the best performing nanofuels we have to deal with many NMs/base fuel pairs that all have to be evaluated with respect to a number of various criteria, ranging from engine break thermal efficiency to possible NMs environmental and human health impacts. Thus, finding the best performing NMs/base fuel pairs is an extremely difficult task, especially, if one takes into account the lack of experimental results data.

To make the situation manageable and to enable getting at least some useful fuel ranking, several compromises have been done in this book as follows.

- First of all, the nanofuels will be evaluated only with respect to the most frequently measured engine characteristics; this means that all other criteria related to, for example, effects on engine parts or NMs health impacts will be neglected.
- Second, only those nanofuels will be taken into consideration for which all of the required data was measured and reported. This means that only a subset of nanofuels given in Tables 5.1, 5.2, 5.3, 5.4, 5.5, 5.6, 5.7, 5.8, 5.9, 5.10, 5.11, 5.12,

and 5.13 will be considered. This set will be termed the *rankable subset* in the rest of this chapter.

The data required for nanofuel evaluation includes: NO_x, HC, CO, and smoke emissions, brake-specific fuel consumption, effective power, and brake thermal efficiency. Note that smoke emission was included here instead of total PM. This is because smoke is the most frequently measured part of the total PM emission, which in contrast is measured rather rarely.

To enable some kind of nanofuel ranking, one needs an appropriate measure. For this purpose, one must define some kind of a *performance factor*, here denoted by ξ, that can be used in determining which fuel performs better. More precisely, the performance factor used here is defined as

$$\xi = \sum_i w_i \psi_\mathrm{i}$$

where $i \in \{\eta_\mathrm{th}, P_\mathrm{e}, g_\mathrm{e}, \mathrm{NO}_x, \mathrm{HC}, \mathrm{CO}, \text{smoke}\}$ denotes a particular engine characteristic and w_i are weighting coefficients, which reflect the relative importance of a particular engine characteristic. Furthermore, the symbols ψ_i denote signed relative variations of engine characteristics, observed when neat base fuel is replaced by the corresponding nanofuel; the term *relative* means that the actually measured variation of some characteristic is divided by the corresponding characteristic value that was obtained with neat base fuel. Note that the weighting factors will be properly signed so that the best performing fuel will exhibit the highest factor value and vice versa.

Unfortunately, due to lack of universally valid and meaningful criteria, the choice of appropriate weighting coefficients is a delicate and problem-dependent task that may influence the obtained result significantly. For this purpose, two sets of weights will be used here to define two variants of the performance factor. By no way, however, these two sets should be considered as the only options possible.

The first performance factor variant, ξ_A, corresponding to the first set of weights, considers all discussed engine characteristics, including engine performances. It is defined by

$$\begin{aligned}\xi_A = 0.05\psi_{\eta_\mathrm{th}} + 0.05\psi_{P_\mathrm{e}} - 0.1\psi_{g_\mathrm{e}} - 0.25\psi_{\mathrm{NO}_x} \\ - 0.15\psi_\mathrm{HC} - 0.15\psi_\mathrm{CO} - 0.25\psi_\mathrm{smoke}\end{aligned}$$

According to the chosen weight values, one can see that among the engine performance characteristics the highest priority was given to brake-specific fuel consumption. Meanwhile, among the emissions, NO_x and smoke got the highest priorities.

The second performance factor variant, ξ_B, does not consider engine performances (the corresponding weights are zero). It is defined as

$$\xi_B = -0.3\psi_{\mathrm{NO}_x} - 0.2\psi_\mathrm{HC} - 0.2\psi_\mathrm{CO} - 0.3\psi_\mathrm{Smoke}$$

where from it follows that obviously the highest priorities were assigned again to NO_x and smoke. Note that in both variants, NO_x and smoke emissions got the highest weighting factors. This reflects the fact that NO_x and PM (including smoke) are considered the most harmful diesel engine emissions with respect to human health.

Both performance factors, ξ_A and ξ_B, were calculated for all nanofuels from the rankable subset. The results, sorted with respect to decreasing performance factor values, are given in Table 5.14. As can be seen, with respect to both performance factors, the most promising nanofuels are Al_2O_3-D/*jj*B20 (case Al-12), followed by Al_2O_3-D/*ts*B20 (case Al-17), ZnO-D/*p*B30/E30 (case Zn-8), MgO-D/W10 (case Mg-1), and Al_2O_3 and CeO_2-*j*B100 (case AlCe-2).

It is evident that the NMs/base fuel pairs, occupying the first quarter of Table 5.14, are the same for both criteria; only the placements of some fuels, ranging from the

Table 5.14 Sorted performance factors for all rankable nanofuels

Order	Variant *A*		Variant *B*		Order	Variant *A*		Variant *B*	
	Case name	ξ_A	Case name	ξ_B		Case name	ξ_A	Case name	ξ_B
1	Al-12	49.2	Al-12	59.5	25	Al-19	13.0	Al-13	14.8
2	Al-17	32.0	Al-17	39.9	26	Ce-9	12.4	Al-10	14.7
3	Zn-8	30.4	Zn-8	36.1	27	Al-13	12.3	Co-3	14.5
4	Mg-1	28.2	Mg-1	34.1	28	Co-3	11.9	Ce-9	14.5
5	AlCe-2	26.8	AlCe-2	32.2	29	Ce-14	11.6	Cu-7	14.2
6	wC-3	24.6	Ti-11	28.4	30	Cu-7	11.5	Ti-5	13.0
7	Ti-11	23.8	C-5	27.5	31	RH-1	11.1	Al-20	12.9
8	C-8	23.4	AlCe-1	27.0	32	Ti-5	10.9	Ce-14	12.7
9	AlCe-1	23.1	wC-3	26.4	43	Al-20	10.9	RH-1	11.5
10	C-5	22.2	C-8	25.7	34	Al-9	10.7	Ce-8	11.4
11	Fe-3	21.9	Ag-4	25.4	35	Ce-8	9.3	Mg-2	10.7
12	Ag-4	20.6	Fe-3	25.3	36	Cu-6	8.8	Al-9	10.7
13	Ti-10	20.4	Ag-5	23.3	37	C-7	7.9	Cu-6	10.1
14	Ag-5	19.7	Ti-10	23.1	38	Ti-6	7.8	C-7	9.5
15	Zn-6	18.5	Zn-6	22.3	39	C-4	7.7	C-4	9.2
16	C-9	18.2	C-9	20.4	40	Al-8	7.7	C-2	8.6
17	Co-1	17.5	Ti-2	19.6	41	Mg-2	7.7	Ti-6	8.4
18	Ti-2	15.9	Co-1	18.9	42	Ce-7	7.5	Al-8	7.8
19	Ce-15	15.5	Ce-15	17.4	43	Al-14	7.3	Ce-7	7.6
20	Al-16	15.3	Al-19	16.7	44	C-2	7.2	Mg-4	7.4
21	Zn-7	14.9	Al-16	15.9	45	C-6	6.0	Al-14	7.1
22	Ce-3	14.0	Ce-3	15.7	46	Ti-4	4.0	C-6	5.1
23	Al-10	13.9	Zn-7	15.5	47	Ti-7	3.6	Ti-4	5.0
24	Fe-4	13.7	Fe-4	15.0	48	Mg-4	3.2	Ti-7	2.1

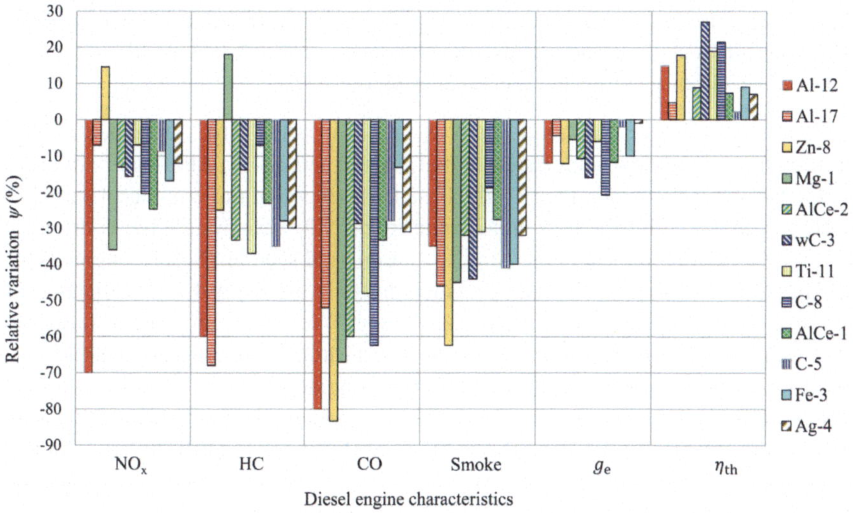

Fig. 5.19 Relative characteristic variations of the best performing NMs-based fuel pairs

6th to 12th place, are different. For a better insight, the relative variation ψ_i of the engine characteristics of these 12 best performing cases is presented in Fig. 5.19. In this graphics, the variation of engine power is not depicted. The reason for this is that in all those cases the engine power variation was practically zero, except in case wcC-3, where the addition of C NTs to *j*B/Bu40 base fuel increased engine power by about 6.7% [31]. Furthermore, it needs to be pointed out that among the subset of rankable nanofuels, listed in Fig. 5.19, only in case Al-12 the engine was tested at several engine operating regimes. So, for this case the characteristic variations were calculated from the results averaged over all tested operating regimes. In all other cases, the variations were obtained from the data measured at only one engine operating regime.

It is evident from Fig. 5.19 that NO_x emission is reduced in all cases, ranked in the first quarter of Table 5.14, the only exception being ZnO NPs when added to the D/*p*B30/E30 base fuel (case Zn-8). In the latter case, the presence of ZnO NPs increases the calorific value and average in-cylinder temperature which leads to higher NO_x emission [93]. In all other best performing cases, NO_x reduction is attributed to:

- heat sink effect of Al_2O_3 NPs in D/*jj*B20 or in D/*ts*B20 base fuel (cases Al-12 and Al-17) [35, 100],
- NO_x reduction agent role of Al_2O_3 and CeO_2 NPs in *j*B100 or in D/*j*B20 base fuel (cases AlCe-1 and AlCe-2) [94, 95] and of TiO_2 NPs in *ci*B100 (case Ti-11) [80] which promotes NO_x conversion into oxygen and N_2 during the combustion process,
- increased cetane number of MgO-D/W10 and Ag_2O-*n*B100 nanofuels by addition of MgO (case Mg-1) and Ag_2O NPs (case Ag-4) which reduces the amount of

fuel burned in the premixed combustion phase; this in turn reduces the in-cylinder temperature and consequently also the NO_x emission produced mainly in the first part of the mixing controlled combustion phase [43].

What is quite interesting about the NO_x emission is the huge difference in NO_x variation of two rather similar cases, namely Al-12 and Al-17. Although both cases involve the same NPs and rather similar base fuels (blend of mineral diesel and 20% biodiesel), they exhibit the best and a rather poor NO_x variation. What is different in both cases is the biodiesel type (jojoba vs. tamarind seed biodiesel), engine type, and operating conditions. The data provided does not allow for the identification of the most influencing reason for this discrepancy. What one can suspect, however, is how sophisticated and difficult it will be to determine optimally everything from nanofuel composition and preparation to proper engine setup.

Regarding the HC emission, by far the largest reductions of more than 60% were obtained by adding Al_2O_3 NPs to D/*ts*B20 (case Al-17) and to D/*jj*B20 (case Al-12). In both cases, this reduction was attributed to the enhanced secondary atomization and catalytic effect of the engaged Al_2O_3 NPs [35, 100].

The largest reduction of CO emission was observed in the cases Zn-8 and Al-12, which involve the ZnO-D/*pb*B30/E30 and Al_2O_3-D/*jj*B20 nanofuels, respectively. For both fuels, this is attributed to the catalytic activity of the involved NPs and their reactivity role of acting as oxidation agents to convert CO into CO_2 during the combustion process [35, 93].

The largest smoke emission reduction of over 40% is achieved by addition of ZnO to D/*pb*B30/E30 (case Zn-8), Al_2O_3NPs to D/*ts*B20 (case Al-17), MgO to MgO-D/W10, and by adding C NTs into *j*B/Bu40 (case C-13). In all these cases, this is attributed to shorter ignition delay, better fuel/air mixing, and catalytic and reactivity effects of the involved NMs [31, 43, 93, 100].

Regarding the economy and performance characteristics, one can see from Fig. 5.19 that the highest improvements of η_{th} and g_e were obtained in cases wC-3 and C-8. The involved mwC and C NTs exhibit high thermal conductivity, and their addition to *j*B/Bu40 and *j*B/*tp*O10 increases the cetane number and calorific value of the fuel. This leads to an efficient combustion process and consequently to an increase of η_{th} and a decrease of g_e[31, 122].

References

1. Aalam CS, Saravanan CG (2017) Effects of nano metal oxide blended Mahua biodiesel on CRDI diesel engine. Ain Shams Eng J 8:689–696
2. Aalam CS, Saravanan CG, Kannan M (2015) Experimental investigations on a CRDI system assisted diesel engine fueled with aluminum oxide nanoparticles blended biodiesel. Alex Eng J 54:351–358
3. Aghbashlo M, Tabatabaei M, Khalife E, Najafi B, Mirsalim SM, Gharehghani A, Mohammadi P, Dadak A, Shojaei TR, Khounani Z (2017) A novel emulsion fuel containing aqueous nano cerium oxide additive in diesel-biodiesel blends to improve diesel engines performance and reduce exhaust emissions: part II—exergetic analysis. Fuel 205:262–271

4. Aghbashlo M, Tabatabaei M, Khalife E, Shojaei TR, Dadak A (2018) Exergoeconomic analysis of a DI diesel engine fueled with diesel/biodiesel (B5) emulsions containing aqueous nano cerium oxide. Energy 149:967–978
5. Ahmad KH, Hossain AK (2017) Impact of nanoparticles and butanol on properties and spray characteristics of waste cooking oil biodiesel and pure rapeseed oil. E3S Web Conf 23:10001. https://doi.org/10.1051/e3sconf/20172310001
6. Anchupogu P, Rao LN, Banavathu B (2018) Effect of alumina nano additives into biodiesel-diesel blends on the combustion performance and emission characteristics of a diesel engine with exhaust gas recirculation. Environ Sci Pollut Res 25:23294–23306
7. Annamalai M, Dhinesh B, Nanthagopal K, Sivaramakrishnan P, Lalvani JIJ, Parthsarathy M, Annamalai K (2016) An assessment on performance, combustion and emission behavior of a diesel engine powered by ceria nanoparticle blended emulsified biofuel. Energy Convers Manag 123:372–380
8. Ashok B, Nanthagopal K, Subbarao R, Johny A, Mohan A, Tamilarasu A (2017) Experimental studies on the effect of metal oxide and antioxidant additives with *Calophyllum inophyllum* Methyl ester in compression ignition engine. J Clean Prod 166:474–484
9. Ashok B, Nanthagopal K, Mohan A, Johny A, Tamilarasu A (2017) Comparative analysis on the effect of zinc oxide and ethanox as additives with biodiesel in CI engine. Energy 140:352–364
10. Ashok B, Nanthagopal K, Vignesh DS (2018) *Calophyllum inophyllum* methyl ester biodiesel blend as an alternate fuel for diesel engine applications. Alex Eng J 57(3):1239–1247
11. Avulapati MM, Ganippa LC, Xia Megaritis A (2016) Puffing and micro-explosion of diesel-biodiesel-ethanol blends. Fuel 166:59–66
12. Avulapati MM, Megaritis T, Xia J, Ganippa L (2019) Experimental understanding on the dynamics of micro-explosion and puffing in ternary emulsion droplets. Fuel 239:1284–1292
13. Basha JS (2018) Impact of carbon nanotubes and di-ethyl ether as additives with biodiesel emulsion fuels in a diesel engine—an experimental investigation. J Energy Inst 91:289–303
14. Basha JS, Anand RB (2011) An experimental investigation in a diesel engine using carbon nanotubes blended water-diesel emulsion fuel. Proc Inst Mech Eng Part A J Power Energy 225:279–288
15. Basha JS, Anand RB (2014) Performance, emission and combustion characteristics of a diesel engine using carbon nanotubes blended Jatropha methyl ester emulsions. Alex Eng J 53:259–273
16. Basu S, Miglani A (2016) Combustion and heat transfer characteristics of nanofluid fuel droplets: a short review. Int J Heat Mass Transf 96:482–503
17. Basu S, Agarwal AK, Mukhopadhyay A, Patel C (2018) Droplets and sprays. Applications for combustion and propulsion. Springer, Singapore
18. Batley GE, Hallburton B, Kirby JK, Doolette CL, Navarro D, McLaughlin MJ, Veitch C (2013) Characterization and ecological risk assessment of nanoparticulate CeO_2 as a diesel fuel catalyst. Environ Toxicol Chem 32:1896–1905
19. Chaichan MT, Kadhum AAH, Al-Amiery AA (2017) Novel technique for enhancement of diesel fuel: impact of aqueous alumina nano-fluid on engine's performance and emissions. Case Stud Therm Eng 10:611–620
20. Chandrasekaran V, Arthanarisamy M, Nachiappan P, Dhanakotti S, Moorthy B (2016) The role of nano additives for biodiesel and diesel blended transportation fuels. Transp Res Part D 46:145–156
21. Chen AF, Adzmi MA, Adam A, Othman MF, Kamaruzzaman MK, Mrwan AG (2018) Combustion characteristics, engine performances and emissions of a diesel engine using nanoparticle-diesel fuel blends with aluminium oxide, carbon nanotubes and silicon oxide. Energy Convers Manag 171:461–477
22. D'Silva R, Binu KG, Thirumaleshwara B (2015) Performance and Emission characteristics of a CI engine fueled with diesel and TiO_2 nanoparticles as fuel additive. Mater Today Proc 2:3728–3735

23. D'Silva R, Hafeez M, Fernandez J, Paloth F, Rahiz IA, Binu KG, Raju K, Thirumaleshwara B (2017) ffect of copperoxide nanoadditives on the performance and emissions characteristics of a CI engine. Energy and Power 7(4):99–104
24. Dale JG, Cox SS, Vance ME, Marr LC, Hochella MF (2017) Transformation of Cerium oxide nanoparticles from a diesel fuel additive during combustion in a diesel engine. Environ Sci Technol 51:1973–1980
25. Damanik N, Ong HC, Tong CW, Mahlia TMI, Silitonga AS (2018) A review on the engine performance and exhaust emission characteristics of diesel engines fueled with biodiesel blends. Environ Sci Pollut Res 25:15307–15325
26. Devarajan Y, Munuswamy DB, Mahalingam A (2018) Influence of nano-additive on performance and emission characteristics of a diesel engine running on neat neem oil biodiesel. Environ Sci Pollut Res 25:26167–26172
27. Devarajan Y, Mahalingam A, Munuswamy DB, Arunkumar T (2018) Combustion, performance, and emission study of a research diesel engine fueled with palm oil biodiesel and its additive. Energy Fuels 32:8447–8452
28. Devarajan Y, Munuswamy DB, Mahalingam A (2019) Investigation on behavior of diesel engine performance, emission, and combustion characteristics using nano-additive in neat biodiesel. Heat Mass Transfer 55(6):1641–1650
29. Dhinesh B, Annamalai M (2018) A study on performance, combustion and emission behaviour of diesel engine powered by novel nano nerium oleander biofuel. J Clean Prod 196:74–83
30. Dhinesh B, Annamalai M, Lalvani IJR, Annamalai K (2017) Studies on the influence of combustion bowl modification for the operation of *Cymbopogon flexuosus* biofuel based diesel blends in a DI diesel engine. Appl Therm Eng 112:627–637
31. El-Seesy AI, Hassan H (2019) Investigation of the effect of adding graphene oxide, graphene nanoplatelet, and multiwalled carbon nanotube additives with n-butanol-Jatropha methyl ester on a diesel engine performance. Renew Energy 132:558–574
32. El-Seesy AI, Abdel-Rahman AK, Bady M, Ookawara S (2016) The influence of multi-walled carbon nanotubes additives into non-edible biodiesel-diesel fuel blend on diesel engine performance and emissions. Energy Proc 100:166–172
33. El-Seesy A, Hassan H, Ookawara S (2018) Effects of graphene nanoplatelet addition to jatropha biodiesel-diesel mixture on the performance and emission characteristics of a diesel engine. Energy 147:1129–1152
34. El-Seesy AI, Hassan H, Ookawara S (2018) Performance, combustion, and emission characteristics of a diesel engine fueled with Jatropha methyl ester and graphene oxide additives. Energy Convers Manag 166:674–686
35. El-Seesy AI, Attia AMA, El-Battsh M (2018) The effect of aluminum oxide nanoparticles addition with Jojoba methyl ester-diesel fuel blend on a diesel engine performance, combustion and emission characteristics. Fuel 224:147–166
36. Ettefaghi E, Ghobadian B, Rashidi A, Najafi G, Khoshtaghaza MH, Rashtchi M, Sadeghian S (2018) A novel bio-nano emulsion fuel based on biodegradable nanoparticles to improve diesel engines performance and reduce exhaust emissions. Renew Energy 125:64–72
37. Fayyazbakhsh A, Pirouzfar V (2016) Determining the optimum conditions for modified diesel fuel combustion considering its emission, properties and engine. Energy Convers Manag 113:209–219
38. Feroskhan M, Ismail S, Gosavi S, Tankhiwale P, Khan Y (2019) Optimization of performance and emissions in a biogas-diesel dual fuel engine with cerium oxide nanoparticle addition. Proc Inst Mech Eng Part D J Automob Eng 233(5):1178–1193
39. Ghafoori M, Ghobadian B, Najafi G, Layeghi M, Rashidi A, Mamat R (2015) Effect of nano-particles on the performance and emission of a diesel engine using biodiesel-diesel blend. Int J Autom Mech Eng 12:3097–3108
40. Guan B, Zhan R, Lin H, Huang Z (2015) Review of the state-of-the-art of exhaust particulate filter technology in internal combustion engines. J Environ Manag 154:225–258
41. Gumus S, Ozcan H, Ozbey M, Topaloglu B (2016) Aluminum oxide and copper oxide nanodiesel fuel properties and usage in a compression ignition engine. Fuel 163:80–87

42. Hajjari M, Ardjmand M, Tabatabaei M (2014) Experimental investigation of the effect of cerium oxide nanoparticles as a combustion-improving additive on biodiesel oxidative stability: mechanism. RSC Adv 4:14252–14356
43. Hasannuddin AK, Yahya WJ, Sarah S, Ithmin AM, Syahrullail S, Sidik NAC, Kassim KAA, Ahmad Y, Hirofumi N, Ahmad MA, Sugeng DA, Zuber MA, Ramlan NA (2018) Nano-additives incorporated water in diesel emulsion fuel: fuel properties, performance and emission characteristics assessment. Energy Convers Manag 169:291–314
44. Heydari-Maleney K, Taghizadeh-Alisaraei A, Ghobadian B, Abbaszadeh-Mayvan A (2017) Analyzing and evaluation of carbon nanotubes additives to diesohol-B2 fuels on performance and emission of diesel engines. Fuel 196:110–123
45. Hoseini SS, Najafi G, Ghobadian B, Mamat R, Sidik NAC, Azmi WH (2017) The effect of combustion management on diesel engine emissions fueled with biodiesel-diesel blends. Renew Sustain Energy Rev 73:307–331
46. Hoseini SS, Najafi G, Ghobadian B, Mamat R, Ebadi MT, Yusaf T (2018) Novel environmentally friendly fuel: The effects of nanographene oxide additives on the performance and emission characteristics of diesel engines fuelled with *Ailanthus altissima* biodiesel. Renew Energy 125:283–294
47. Hosseini SH, Taghizadeh-Alisaraei A, Ghobadian B, Abbaszadeh-Mayvan A (2017) Effect of added alumina as nano-catalyst to diesel-biodiesel blends on performance and emission characteristics of CI engine. Energy 124:543–552
48. Hosseini SH, Taghizadeh-Alisaraei A, Ghobadian B, Abbaszadeh-Mayvan A (2017) Performance and emission characteristics of a CI engine fuelled with carbon nanotubes and diesel-biodiesel blends. Renew Energy 111:201–213
49. Hosseinzadeh-Bandbafha H, Tabatabaei M, Aghbashlo M, Khanali M, Demirnas A (2018) A comprehensive review on the environmental impacts of diesel/biodiesel additives. Energy Convers Manag 174:579–614
50. Imdadul HK, Masjuki HH, Kalam MA, Zulkifli NWM, Rashed MM, Rashedul HK, Monirul IM, Mosarof MH (2015) A comprehensive review on the assessment of fuel additive effects on combustion behavior in CI engine fuelled with diesel biodiesel blends. RSC Adv 5:67541–67567
51. Ismael MA, Heikal MR, Aziz ARA, Syah F, Zainal EZA, Crua C (2018) The effect of fuel injection equipment on the dispersed phase of water-in-diesel emulsions. Appl Energy 222:762–771
52. Jayanthi P, Srinivasa RM (2016) Effect of nanoparticles additives on performance and emissions characteristics of a DI diesel engine fueled with biodiesel. Int J Adv Eng Technol 9(6):689–695
53. Jeryrajkumar L, Anbarasu G, Elangovan T (2016) Effects on nano additives on performance and emission characteristics of *Calophyllum inophyllum* biodiesel. Int J Chem Tech Res 9(4):210–219
54. Jiang L, Wang YD, Roskilly AP, Xie XL, Zhang ZC, Wang RZ (2018) Investigation on thermal properties of a novel fuel blend and its diesel engine performance. Energy Convers Manag 171:1540–1548
55. Jiaqiang E, Zhang Z, Chen J, Pham M, Zhao X, Peng Q, Zhang B, Yin Z (2018) Performance and emission evaluation of a marine diesel engine fueled by water biodiesel-diesel emulsion blends with a fuel additive of a cerium oxide nanoparticle. Energy Convers Manag 169:194–205
56. Kamesh MR, Madhu D (2017) Effect of nanoparticles on the emissions of a CI engine. Int J Renew Energy Res 7(2):751–757
57. Kannaiyan K, Sadr R (2017) The effects of alumina nanoparticles as fuel additives on the spray characteristics of gas-to-liquid jet fuels. Exp Thermal Fluid Sci 87:93–103
58. Kannaiyan K, Anoop K, Sadr R (2017) Effects of nanoparticles on the fuel properties and spray performance of aviation turbine fuel. J Energy Res Technol 139(3):1–8
59. Karthikeyan S, Prathima A (2017) Environmental effect of CI engine using microalgae methyl ester with doped nano additives. Transp Res Part D 50:385–396

60. Karthikeyan S, Elango A, Silaimani SM, Prathima A (2014) Role of Al_2O_3 nano additive in GSO biodiesel on the working characteristics of a CI engine. Indian J Chem Technol 21:285–289
61. Karthikeyan S, Elango A, Prathima A, Raja K (2014) Environmental effects of nano additive Co_3O_4 in Grape seed oil biofuel fueled in CI engine. Res J Chem Environ 18(5):14–18
62. Kegl B, Kegl M, Pehan S (2013) Green diesel engines. Biodiesel usage in diesel engines. Springer, London
63. Khalife E, Tabatabaei M, Demirbas A, Aghbashlo M (2017) Impacts of additives on performance and emission characteristics of diesel engines during steady state operation. Prog Energy Combust Sci 59:32–78
64. Khalife E, Tabatabei M, Najafi B, Mirsalim SM, Gharehghani A, Mohammadi P, Aghbashlo M, Ghaffari A, Khounani Z, Shojaei TR, Salleh MAM (2017) A novel emulsion fuel containing aqueous nano cerium oxide additive in diesel-biodiesel blends to improve diesel engines performance and reduce exhaust emissions: part I—experimental analysis. Fuel 741–750
65. Khond VW, Kriplani VM (2016) Effect of nanofluid additives on performances and emissions of emulsified diesel and biodiesel fueled stationary CI engine: a comprehensive review. Renew Sustain Energy Rev 59:1338–1348
66. Kumar S, Dinesha P, Bran I (2017) Influence of nanoparticles on the performance and emission characteristics of a biodiesel fuelled engine: an experimental analysis. Energy 140:98–105
67. Kumar MV, Babu AV, Kumar PP (2018) The impacts on combustion, performance and emissions of biodiesel by using additives in direct injection diesel engine. Alex Eng J 57:509–516
68. Kumar MV, Babu AV, Kumar PR (2019) Influence of metal-based cerium oxide nanoparticle additive on performance, combustion, and emissions with biodiesel in diesel engine. Environ Sci Pollut Res 26:7651–7664
69. Lapuerta M, Rodríguez-Fernández J, Fernández-Rodríguez D, Patiño-Camino R (2017) Modeling viscosity of butanol and ethanol blends with diesel and biodiesel fuels. Fuel 199:332–338
70. Lawrence KR, Tharigonda H, Alluru G (2018) An experimental investigation on combined effect of EGR and SCR for diesel engine fueled with the addition of TiO_2 nanoparticle to *M. Elangi* methyl ester blend. Aust J Mech Eng https://doi.org/10.1080/14484846.2018.1519053
71. Logothetidis S, Patsalas P, Charitidis C (2003) Enhanced catalytic activity of nanostructures cerium oxide films. Mater Sci Eng C 23:803–806
72. Mehregan M, Moghiman M (2018) Effects of nano-additives on pollutants emission and engine performance in urea-SCR equipped engine fueled with blended-biodiesel. Fuel 222:402–406
73. Mehta RN, Chakraborty M, Parikh PA (2014) Nanofuels: combustion, engine performance and emissions. Fuel 120:91–97
74. Mingle K, Lauterbach J (2018) Synthesis-structure-activity relationships in Co_3O_4 catalyzed CO oxidation. Front Chem 6:185
75. Mirzajanzadeh M, Tabatabaei M, Ardjmand M, Rashidi A, Ghobadian B, Barkhi M, Pazouki M (2015) A novel soluble nano-catalysts in diesel-biodiesel fuel blends to improve diesel engines performance and reduce exhaust emissions. Fuel 139:374–382
76. Mohankumar S, Senthilkumar P (2017) Particulate matter formation and its control methodologies for diesel engine: a comprehensive review. Renew Sustain Energy Rev 80:1227–1238
77. Muthusamy S, Nallathambi SS, Ramasamy RK, Mohamed ST (2018) Effects of nanoparticles blended biodiesel on single cylinder CI engine. Mater Today Proc 5:6831–6838
78. Naik V, Kumar KK (2018) Performance and emission characteristics of diesel engines with Al_2O_3 and CuO nanoparticles as additives. Int J Mech Eng Technol 9(2):791–798
79. Najafi G (2018) Diesel engine combustion characteristics using nano-particles in biodiesel-diesel blends. Fuel 212:668–678
80. Nanthagopal K, Ashok B, Tamilarasu A, Johny A, Mohan A (2017) Influence on the effect of zinc oxide and titanium dioxide nanoparticles as an additive with *Calophyllum inophyllum* methyl ester in a CI engine. Energy Convers Manag 146:8–19

81. Nayak S, Shet VB, Rao CV, Joshi K (2018) Performance evaluation and emission characteristics of a 4 stroke diesel engine using green synthesized silver nanoparticles blended biodiesel. Mater Today Proc 5:7889–7897
82. Nour M, El-Seesy AI, Abdel-Rahman AK, Bady M (2018) Influence of adding aluminum oxide nanoparticles to diesterol blends on the combustion and exhaust emission characteristics of a diesel engine. Exp Thermal Fluid Sci 98:634–644
83. Ooi JB, Ismail HM, Swamy V, Wang X, Swain AK, Rajanren JR (2016) Graphite oxide nanoparticle as a diesel fuel additive for cleaner emissions and lower fuel consumption. Energy Fuels 230:1341–1353
84. Ooi JB, Ismail HM, Tan BT, Wang X (2018) Effects of graphite oxide and single-walled carbon nanotubes as diesel additives on the performance, combustion, and emission characteristics of a light-duty diesel engine. Energy 161:70–80
85. Örs I, Sarıkoç S, Atabani AE, Ünalan S, Akansu SO (2018) The effects on performance, combustion and emission characteristics of DICI engine fuelled with TiO_2 nanoparticles addition in diesel/biodiesel/n-butanol blends. Fuel 234:177–188
86. Özgür T, Özcanli M, Aydin K (2015) Investigation of nanoparticle additives to biodiesel for improvement of the performance and exhaust emissions in a compression ignition engine. Int J Green Energy 12:51–56
87. Pandian AK, Ramakrishnan RBB, Devarajan Y (2017) Emission analysis on the effect of nanoparticles on neat biodiesel in unmodified diesel engine. Environ Sci Pollut Res 24:23273–23278
88. Panithasan MS, Gopalakichenin D, Venkadesa G, Veeraraagavan S (2018) Impact of rice husk nanoparticle on the performance and emission aspects of a diesel engine running on blends of pine oil-diesel. Environ Sci Pollut Res 26(1):282–291
89. Paramashivaih BM, Banapurmath NR, Rajashekhar CR, Khandal SV (2018) Studies on effect of graphene nanoparticles addition in different levels with *Simarouba* biodiesel and diesel blends on performance, combustion and emission characteristics of CI engine. Arab J Sci Eng 43:4793–4801
90. Patel HK, Kumar S (2017) Experimental analysis on performance of diesel engine using mixture of diesel and bio-diesel as a working fuel with aluminum oxide nanoparticle additive. Therm Sci Eng Progress 4:252–258
91. Patil DS, Chopade DA, Kumbhalkar MA (2018) Experimental investigation of effect of cerium oxide nanoparticles as a fuel additive in cottonseed biodiesel blends. MAYFEB J Mech Eng 1:1–12
92. Perumal V, Ilangkumaran M (2018) The influence of copper oxide nano particle added pongamia methyl ester biodiesel on the performance, combustion and emission of a diesel engine. Fuel 232:791–802
93. Prabakaran B, Udhoji A (2016) Experimental investigation into effects of addition of zinc oxide on performance, combustion and emission characteristics of diesel-biodiesel-ethanol blends in CI engine. Alex Eng J 55:3355–3362
94. Prabu A (2017) Nanoparticles as additive in biodiesel on the working characteristics of a DI diesel engine. Ain Shams Eng J 9(4):2343–2349
95. Prabu A (2018) Engine characteristic studies by application of antioxidants and nanoparticles as additives in biodiesel diesel blends. J Energy Res Technol 140:1–7
96. Prakash PJ, Reddy SSK (2016) A review study of performance and emissions of CI engine fuelled with blend of biodiesel along with fuel nano additives. Int Res J Eng Technol 3(6):2356–2359
97. Praveen A, Rao GLN, Balakrishna B (2018) Performance and emission characteristics of a diesel engine using *Calophyllum inophyllum* biodiesel blends with TiO_2 nanoadditives and EGR. Egypt J Petrol 27:731–738
98. Praveena V, Narayanan SA, Hariskousik G, Balavinayagam K (2018) Effect of aluminium hydroxide nano fluid in castor oil biodiesel fueled diesel engine. Mater Sci Eng 402:012036
99. Radhakrishnan S, Munuswamy DB, Devarajan Y, Arunkumar T, Mahalingam A (2018) Effect of nanoparticle on emission and performance characteristics of a diesel engine fueled with cashew nut shell biodiesel. Energy Sources A Recov Util Environ Effects 40(20):2485–2493

100. Raju VD, Kishore PS, Nanthagopal K, Ashok B (2018) An experimental study on the effect of nanoparticles with novel tamarind seed methyl ester for diesel engine applications. Energy Convers Manag 64:655–666
101. Ramanan MV, Yuvarajan D (2016) Emission analysis on the influence of magnetite nanofluid on methyl ester in diesel engine. Atmosp Pollut Res 7:477–481
102. Ramarao K, Rao CJ, Sreeramulu D (2015) The experimental investigation on performance and emission characteristics of a single cylinder diesel engine using nano additives in diesel and biodiesel. Indian J Sci Technol 8(29):75718
103. Ramesh DK, Dhananjaya KJL, Hemanth KSG, Namith V, Parashuram BJ, Sharath S (2018) Study on effects of Alumina nanoparticles as additive with Poultry litter biodiesel on performance, combustion and emission characteristic of diesel engine. Mater Today Proc 5:1114–1120
104. Rani AS, Ajaad D, Karteek P, Naidu AL (2017) Experimental investigation on the performance and emission characteristics of di-diesel engine using diesel-ethanol blends and aluminium oxide nano particles. Int J Mech Product Eng Res Dev 7(5):301–310
105. Rani GJ, Rao YVH, Balakrishna B (2018) Investigation of fuel injection pressure impact on CI engine performance and emissions using biodiesel blend with alumina nano additives. Int J Mech Eng Technol 9(6):922–928
106. Ranjan A, Dawn SS, Jayaprabakar J, Nirmala N, Saikiran K, Sriram SS (2018) Experimental investigation on effect of MgO nanoparticles on cold flow properties, performance, emission and combustion characteristics of waste cooking oil biodiesel. Fuel 220:780–791
107. Rao MS, Anand RB (2016) Performance and emission characteristics improvement studies on a biodiesel fuelled DICI engine using water and AlO(OH) nanoparticles. Appl Therm Eng 98:636–645
108. Sabarish R, Mahankumar D, Kumar MPJ, Manavalan S (2018) Experimental study of nano additive with biodiesel and its blends for diesel engine. Int J Pure Appl Math 118(18):967–979
109. Sahoo RR, Jain A (2019) Experimental analysis of nanofuel additives with magnetic fuel conditioning for diesel engine performance and emissions. Fuel 236:365–372
110. Sajeevan AC, Sajith V (2013) Diesel engine emission reduction using catalytic nanoparticles: an experimental investigation. J Eng 589382:1–9
111. Sajith V, Sobhan CB, Peterson GP (2010) Experimental investigations on the effects of cerium oxide nanoparticle fuel additives on biodiesel. Adv Mech Eng 581407:1–6
112. Samec N, Kegl B, Dibble RW (2002) Numerical and experimental study of water/oil emulsified fuel combustion in a diesel engine. Fuel 81:2035–2044
113. Sandeep K, Rajashekhar CR, Karthik SR (2018) Experimental studies on effect of nano particle blended biodiesel combustion on performance and emission of CI engine. Mater Sci Eng 376:012019
114. Saraee HS, Jafarmadar S, Taghavifar H, Ashrafi SJ (2015) Reduction of emissions and fuel consumption in a compression ignition engine using nanoparticles. Int J Environ Sci Technol 12:2245–2252
115. Saraee HS, Taghavifar H, Jafarmadar S (2017) Experimental and numerical consideration of the effect of CeO_2 nanoparticles on diesel engine performance and exhaust emission with the aid of artificial neural network. Appl Therm Eng 113:663–672
116. Saxena V, Kumar N, Saxena VK (2017) A comprehensive review on combustion and stability aspects of metal nanoparticles and its additive effect on diesel and biodiesel fueled CI engine. Renew Sustain Energy Rev 70:563–588
117. Sazhin SS, Rybdylova O, Crua C, Heikal M, Ismael MA, Nissar Z, Aziz ARBA (2019) A simple model for puffing/micro-explosions in water-fuel emulsion droplets. Int J Heat Mass Transf 131:815–821
118. Seela CR, Ravisankar B, Raju BMVA (2017) A GRNN based frame work to test the influence of nano zinc additive biodiesel blends on CI engine performance and emissions. Egypt J Petrol 27(4):641–647
119. Shaafi T, Velraj R (2015) Influence of alumina nanoparticles, ethanol and isopropanol blend as additive with diesel-soybean biodiesel blend fuel: Combustion, engine performance and emissions. Renew Energy 80:655–663

120. Sharma CK, Sharma M, Verma O, Sharma V (2015) Green synthesis of different nanoparticles and their potential applications in different fields—a critical review. Int J Pharm Bio Sci 6(3):555–567
121. Sharma SKS, Das RK, Sharma A (2016) Improvement in the performance and emission characteristics of diesel engine fueled with jatropha methyl ester and tyre pyrolysis oil by addition of nano additives. J Brazil Soc Mech Sci Eng 38:1907–1920
122. Sharma AK, Sharma A, Ahmad N, Khan W (2016) Zinc oxide nanoparticle fuel additives for improved efficiency and emissions of internal combustion engines. Adv Sci Eng Med 8:1–5
123. Sim HS, Yetter RA, Hong S, Duin ACT, Dabbs DM, Aksay IA (2020) Functionalized graphene sheet as a dispersible fuel additive for catalytic decomposition of methylcyclohexane. Combust Flame 217:212–221
124. Singh AS, Madras G (2015) Detailed mechanism and kinetic study of CO oxidation on cobalt oxide surfaces. Appl Catal A Gen 504:463–475
125. Sivakumar M, Sundaram NS, Kumar RR, Thasthagir MHS (2018) Effect of aluminium oxide nanoparticles blended pongamia methyl ester on performance, combustion and emission characteristics of diesel engine. Renew Energy 116:518–526
126. Soudagar MEM, Nik-Ghazali N-N, Kalam MdA, Badruddin IA, Banapurmath NR, Akram N (2018) The effect of nano-additives in diesel-biodiesel fuel blends: a comprehensive review on stability, engine performance and emission characteristics. Energy Convers Manag 178:146–177
127. Srinidhi C, Madhusudhan A (2017) A diesel engine performance investigation fuelled with nickel oxide nano fuel-methyl ester. Int J Renew Energy Res 7(2):676–681
128. Srinidhi C, Madhusudhan A, Channapattana SV (2019) Effect of NiO nanoparticles on performance and emission characteristics at various injection timings using biodiesel-diesel blends. Fuel 235:185–193
129. Sundaram D, Yang V, Yetter RA (2017) Metal-based nanoernergetic materials: Synthesis, properties, and applications. Prog Energy Combust Sci 61:293–365
130. Sundararajan NK, Ammal ARB (2018) Improvement studies on emission and combustion characteristics of DICI engine fuelled with colloidal emulsion of diesel distillate of plastic oil, TiO_2 nanoparticles and water. Environ Sci Pollut Res 25:11595–11613
131. Vairamuthu G, Sundarapandian S, Kailasasanathan C, Thangagiri B (2016) Experimental investigation on the effects of cerium oxide nanoparticle on *Calophyllum inophyllum* (Punnai) biodiesel blended with diesel fuel in DI diesel engine modified by nozzle geometry. J Energy Inst 89:668–682
132. Vellaiyan S (2019) Enhancement in combustion, performance, and emission characteristics of a diesel engine fueled with diesel, biodiesel, and its blends by using nanoadditive. Environ Sci Pollut Res 26(10):9561–9573
133. Venkatesan H, Sivamani S, Sampath S, Gopi V, Kumar DM (2017) A comprehensive review on the effect of nano metallic additives on fuel properties, engine performance and emission characteristics. Int J Renew Energy Res 7(2):825–843
134. Venu H, Madhavan V (2016) Effect of Al_2O_3 nanoparticles in biodiesel-diesel-ethanol blends at various injection strategies: performance, combustion and emission characteristics. Fuel 186:176–189
135. Venu H, Raju VD, Subramani L (2019) Combined combined effect of influence of nano additives, combustion chamber geometry and injection timing in a DI diesel engine fuelled with ternary (diesel-biodiesel-ethanol) blends. Energy 174:386–406
136. Vinukumar K, Azhagurajan A, Vettivel SC, Vedaraman N, Lenin AH (2018) Biodiesel with nano additives from coconut shell for decreasing emissions in diesel engines. Fuel 222:180–184
137. Vinukumar K, Azhagurajan A, Vettivel SC, Vedaraman N, Lenin AH (2018) Rice husk as nanoadditive in diesel-biodiesel fuel blends used in diesel engine. J Therm Anal Calorim 131:1333–1343
138. Wang Z, Wu S, Huang Y, Huang S, Shi S, Cheng X, Huang R (2018) Experimental investigation on spray, evaporation and combustion characteristics of ethanol-diesel, water-emulsified diesel and neat diesel fuels. Fuel 231:438–448

139. Wu Q, Xie X, Wang Y, Roskilly T (2018) Effect of carbon coated aluminum nanoparticles as additive to biodiesel-diesel blends on performance and emission characteristics of diesel engine. Appl Energy 221:597–604
140. Yuvarajan D, Babu MD, Kumar NB, Kishore PA (2018) Experimental investigation on the influence of titanium dioxide nanofluid on emission pattern of biodiesel in a diesel engine. Atmosp Pollut Res 9:47–52
141. Zhang ZH, Balasubramanian R (2015) Influence of an iron-based fuel-borne catalyst on physicochemical and toxicological characteristics of particulate emissions from a diesel engine. Appl Energy 146:270–278
142. Zhang ZH, Balasubramanian R (2017) Effects od cerium oxide and ferrocene nanoparticles addition as fuel-borne catalyst on diesel engine particulate emissions: environmental and health implications. Environ Sci Technol 51:4248–4258
143. Zhang Y, Huang R, Huang Y, Huang S, Zhou P, Chen X, Qin T (2018) Experimental study on combustion characteristics of an n-butanol-biodiesel droplet. Energy 160:490–499
144. Zhang X, Li T, Wang B, Wei Y (2018) Superheat limit and micro-explosion in droplets of hydrous ethanol-diesel emulsions at atmospheric pressure and diesel-like conditions. Energy 154:535–543

Chapter 6
Practical Viability of Nanofuels Usage in Diesel Engines

As shown in Chap. 5, laboratory experiments confirm that nanofuels may offer a lot of important benefits that are practically immediately visible or measurable via various engine characteristics. Therefore, one can say that these benefits as well as the drawbacks observed during those experiments can roughly be classified as *short-term* consequences of nanofuels usage in diesel engines.

Once a technical novelty becomes attractive, however, it becomes unavoidable to look carefully for its viability in long-term daily use. Within this context, practical technical issues have to be resolved and *long-term* consequences have to be analyzed or at least estimated as good as possible. This chapter addresses these issues in relation to nanofuels usage in a diesel engine. The discussion presented is based on the most recent literature, which, unfortunately, is rather modest. Therefore, one can say right away that practical viability evaluation of nanofuels usage has still a long way to go.

6.1 Technical Issues and Challenges

Nanofuels came into focus rather recently and, since current intensive research is mainly related to engine characteristics, the literature about the practical technical aspects of their usage in diesel engines is rather modest. Therefore, to identify at least the main technical issues and their related challenges, a careful examination of published work and some indirect inference is needed. By using such an approach, one can identify technical issues that may be classified into four groups related to:

- NMs size and dosage,
- nanofuels stability,
- DPF efficiency and durability, and
- ECU programming.

T. Kegl et al., *Nanomaterials for Environmental Application*, Green Energy and Technology, https://doi.org/10.1007/978-3-030-54708-0_6

To make a nanofuel viable, all of the identified technical issues have to be addressed and resolved somehow by economically acceptable solutions, Table 6.1.

6.1.1 NMs Size and Dosage

From laboratory testing done so far, it follows that engine characteristics depend heavily on the NMs size and dosage used with a particular base fuel. In this scope, NMs morphology also seems to be relevant, but the related data in the literature is scarce making some useful conclusions practically impossible. Thus, by putting morphology aside, one is left with two important nanofuel parameters, which must be matched optimally to a number of other parameters such as the type of the base fuel and engine operating regime [39, 49]. Thus, determining the optimal NMs size and dosage for all relevant situations will require extensive laboratory testing and numerical simulations. Although this will surely be a difficult and costly undertaken, there is probably no way around. Namely, inadequate NMs size and dosage may not only degrade performance but also cause problems related to engine operation. For example, in the case of exceeded NMs limit dosage, Kumar et al. [20] observed high level of knocking and unstable engine operation; in this particular case, such behavior was attributed to increased time needed for the nanofuel to mix with air and get ignited.

It should be pretty obvious that optimal NMs dosage depends on the used base fuel. However, what makes this situation far more complex is that the dosage to be optimal also has to be adapted to a particular engine operating regime [7, 30]. The variations needed are by no means insignificant, as it was clearly shown by Nour et al. [30]. They tested various dosages of Al_2O_3 NPs added to a blend of mineral diesel, jojoba biodiesel, and ethanol; the NMs dosage varied in the range from 25 to 100 mg/L. It was clearly shown that the maximum brake thermal efficiency at medium engine loads is reached at 25 mg/L while a dosage of 75 mg/L turned out to be optimal at high loads. El-Seesy et al. [7] run tests with a blend of mineral diesel and jojoba biodiesel used as base fuel. The conclusions were essentially similar in the sense that various dosages of Al_2O_3 NPs were needed in order to obtain maximal reductions of harmful emissions and best engine performances at various engine loads and speeds.

In practical application, variable NMs dosing could be achieved by engaging a suitable device which would prepare the nanofuel with proper dosage just before it is delivered into the engine cylinder. Determining the proper dosage could be controlled by the ECU, which, in turn, would need to be equipped by adequate maps. Unfortunately, to create these ECU maps, extensive and long-term laboratory testing will be needed to determine a viable compromise regarding all aspects of diesel engine operation.

NMs size is obviously a critical parameter that determines NMs properties and activities. Optimal size is therefore crucial in order to get improved engine characteristics. Apart from engine characteristics, however, NMs transformations during

Table 6.1 Overview of practical technical issues and the associated challenges

	Technical issues	Challenges
NMs size and dosage	• Synthesis of NMs of appropriate size • Ensuring optimal NMs size in dependence on actually used base fuel • Ensuring optimal NMs dosage in dependence on nanofuel type and current engine operating regime	• Development of suitable NMs synthesis methods • Extensive laboratory testing of various NMs sizes and dosages in various base fuels at all important engine operating regimes; development of adequate numerical simulation models • Development of subsystems for controlled NMs addition and mixing of NMs and base fuel with respect to various engine operating regimes
Nanofuels stability	• Preparation of nanofuels that are free of NMs agglomeration and exhibit long-term stability	• Long-term testing of stability of various nanofuels • Determination of suitable methods for particular nanofuel preparation with/without surfactants • Development of special systems to prevent deposition/sedimentation of NMs in the tank, tubes, etc.
DPF efficiency and durability	• Prevention of NMs release into the atmosphere • Prevention of DPF clogging • Ensuring DPF durability	• Long-term testing to determine the influence of NMs on DPF • Determination of required DPF modifications (type, material, pore size, wall thickness, cell design and density, filter volumetric capacity) • Development of additional systems to capture and remove NMs particles from the exhaust gas
ECU programming	• Electronically controlled NMs dosage in dependence on current engine operating regime • Electronically controlled values of injection parameters in dependence on used nanofuel and current engine operating regime	• Extensive laboratory testing of NMs dosages in various base fuels at all important engine operating regimes; development of adequate numerical simulation models • Optimized ECU maps of injection and combustion parameters for optimal NMs dosage in dependence on used nanofuel

their travel through the engine and exhaust system as well as the NMs ultimate fate in the environment also depend on the size [5, 17, 28, 40, 47]. These facts make the determination of the optimal size an even more challenging task. On top of that, one has to realize that NMs production methods usually do not allow to be simply tuned to deliver any desired NMs size and morphology. Thus, getting the really optimal NMs size might also require to find a suitable production method. Finally, even if all these challenges are resolved, one faces again the problem of adjusting the optimal NMs size to current engine operating regime. Unfortunately, this would be technically rather difficult to fulfill because this would require to have various NMs sizes stored on board and a special device in order to prepare the nanofuel as needed.

6.1.2 Nanofuels Stability

Nanofuel stability issues would be of minor importance, if the nanofuel would be prepared on board just before being delivered into the engine cylinder for combustion, for example, by a device which would also take care for optimal MNs dosage. However, stability may become a serious problem if the nanofuel is prepared in advance and stored in the fuel tank for later use. In this case, the NMs dispersed in the base fuel may start to agglomerate. Independently of this, NMs sedimentation may also occur. Although sedimentation may occur without agglomeration, it usually becomes worse once NMs start to agglomerate due to their high surface activity. At this point, it might be worth noting that, among other things, agglomeration causes variations in thermo-physical properties of nanofuels like thermal conductivity, viscosity, and density [6, 40].

Long-term stability depends on several parameters with NMs size and nanofuel preparation method being among the most exposed ones. Regarding the size, the smaller particles have a better surface/volume ratio; this reduces the sedimentation effects and effectively makes the fuel more stable [38, 49]. The most important parameter, however, seems to be the nanofuel preparation method. As can be seen in Chap. 4.1, quite some nanofuel preparation methods are being used. Either various methods or similar methods with various process parameters may result in quite different nanofuel stabilities. On top of this, the used surfactant may also have a very important role.

Anyhow, if the nanofuel will be prepared in advance and stored in the fuel tank, it has to exhibit adequate stability. To ensure this, one would need to:

- determine the most suitable method for a particular nanofuel preparation, including optimal process parameters, and surfactants, and
- if necessary, engage special devices to prevent sedimentation of NMs in the fuel tank and tubes.

Whatever steps will be needed, long-term stability of nanofuels to be stored for later use is one of the critical issues that must be addressed and resolved to ensure practical viability of nanofuels usage in diesel engines [6, 40].

6.1.3 DPF Efficiency and Durability

So far, laboratory experimental work related to the effects of nanofuels usage on DPF is rather scarce. In spite of this, it is pretty obvious that the nanofuel effects on the activity, efficiency, and durability of the FPF cannot be neglected [8, 14, 31, 48]. Under some special circumstances, the presence of NMs can even bring some benefits, for example, when a particular NMs have the ability to increase the DPF catalytic activity by effectively increasing the catalytically active surface [14, 42]. Mostly, however, the presence of NMs within the DPF comes with new challenges or inconveniences.

One of the possible drawbacks of NMs presence is increased ash accumulation in the DPF. This can affect the performance of the whole after-treatment system and consequently all diesel engine characteristics [48]. Another important issue is related to DPF filtration capability, which is very good for particulate sizes above 50 nm; for particulates below this limit the DPF efficiency may drop significantly [5, 31]. Unfortunately, NMs are typically sized below this limit [40, 48], meaning that the NMs presence has to be investigated with respect to filtration efficiency. Consequently, the DPF capture and storage capacity with regard to all exhaust particulates also has to be investigated since all these particulates have to be eliminated or burned off somehow to regenerate the DPF.

Once there is more insight into these topics, this will enable the development of new and better adapted DPFs. The most important DPF parameters to be addressed are pore diameter, wall thickness, cell design, cell density, and volumetric capacity [48]. If this is done well, it is supposed that notable gains are possible related to DPF efficiency and durability, and consequently, to PM emission.

Apart from improving the DPF, another way to mitigate problems related to NMs usage, is the introduction of a new device for capturing and removing NMs particles. The location of such a device could be determined based on the trade-off between possible NMs benefits and problems caused in the DPF. If a particular NMs proves to be useful, for example, by enhancing the catalytic activity of the DPF or by reducing the soot ignition temperature and thus helping to regenerate the filter, such a device could be placed after the DPF. Otherwise, the specialized capture device placement before the DPF could be more useful. Besides helping the DPF, such a specialized NMs capturing device could also help enormously in minimizing NMs release into the environment.

Unfortunately, the development of a NMs capturing device might prove quite difficult. This is because, firstly, this device must exhibit minimal influence on temperature, pressure, and transport velocities in the diesel engine exhaust system. And, secondly, once the NMs particles are captured, they would need to be removed somehow and possibly stored for later disposal or recycling. Of course, NMs particles capturing is by itself a challenge. In this view, magnetic particles have an important advantage against the nonmagnetic ones. The obvious reason is that magnetic particles can be much easier collected from the exhaust [21], by engaging permanent magnets. This offers a good opportunity to build an efficient capturing device, although its practical implementation may still not be easy.

6.1.4 ECU Programming

The ECU is used to set the injection parameters according to current engine operating conditions such as engine load, speed, temperature, and so on. This is achieved via the so-called ECU maps that have to be determined carefully in order to achieve optimal engine operation under various conditions.

For a diesel engine, the ECU maps are normally defined for mineral diesel. If the fuel supplied to the engine is changed, however, the ECU maps must also change. Since NMs additives may cause significant variations in fuel properties [19, 38, 40], this means that engagement of a nanofuel would unavoidably require an adequate update of the ECU maps. Unfortunately, even if one selects a particular nanofuel, its properties may still vary notably by varying the NMs dosage. Thus, if the engine would be equipped with a device for variable NMs dosing, the ECU maps would need to depend also on NMs dosage. This would be an additional ECU input parameter and the laboratory testing needed for ECU maps determination would become even more extensive.

By nature, the ECU maps determination process is plagued by trade-offs between many parameters ranging from engine performance to various ecology characteristics. In this context, numerical optimization might offer a useful tool, provided that numerical simulation models with satisfactory accuracy are available. An attempt to engage numerical optimization was reported by Vellaiyan et al. [46]. The objective was to determine optimal content of water and soybean biodiesel in mineral diesel as well as optimal ZnO NPs dosage. These objectives were translated into requests to maximize brake thermal efficiency and minimize specific fuel consumption, HC, CO, and NO_x emissions, and smoke opacity. Although this is by definition a multi-objective optimization problem, a simplified approach was adopted by which all the maximized or minimized quantities were combined into a single-objective function. For this purpose, appropriate weighting factors were introduced to scale adequately the relative importance of individual quantities. Note that a conversion of a multi-objective optimization problem into a single objective one will always simplify the situation from the mathematical point of view. From engineering point of view, however, this means that one has to select proper weighting factors. This is by no way an easy and straightforward task because these factors influence heavily the result; and in most situations, one often has no physically meaningful indicators how to define them.

6.2 Environmental and Human Health Risks

Some of the most important motives for the introduction of NMs, serving as fuel additives, are related to environmental concerns. Namely, as the laboratory tests confirm, the NMs indeed have a potential to reduce various harmful diesel engine emissions, such as NO_x, HC, and CO. Unfortunately, the outlook for the PM emission is somewhat more grim. Although NMs fuel additives have the ability to improve at least some aspects of urban air quality and mitigate global warming problems [43], they practically unavoidably increase the PM count emission. Moreover, under unfavorable circumstances, NMs additives may even facilitate the formation of new self-nucleated metallic particles [48, 50, 51]. This is a reason for a serious concern since NMs can have quite harmful effects on the environment and human health.

According to the prospect of increased PM count emission, the capabilities of the DPF and an optional specialized NMs capture device become increasingly important. Unfortunately, laboratory tests have shown that although high-efficiency DPFs can do a really good work, some amount of NMs is always emitted into the environment together with the engine exhaust [3, 12]. The NMs emitted directly into the environment, even in small amounts, are surely a reason for concern. Apart from this, however, there is another concerning issue. Namely, the NMs accumulated in the DPF are also very problematic if the DPF is not disposed or recycled properly. In such a case, for example, if the DPFs are disposed at unsuitable deposits, the NMs may sooner or later also end up in natural environment [12].

6.2.1 Environment

NMs from nanofuels enter the natural environment from two sources: engine exhaust and improperly disposed DPFs or NMs capture devices. NMs from engine exhaust are released into the atmosphere, wherefrom they are gradually removed by atmospheric precipitation, sedimentation in the form of dust, and coagulation with natural aerosol particles. Therefore, airborne NMs end up sooner or later in terrestrial and aquatic environments since NMs do not evaporate and there is no diffusive NMs transport in the opposite direction [26]. Terrestrial and aquatic environments are also the primary destinations for the NMs from improperly disposed DPFs.

Once the NMs enter the natural environment, they often begin to interact with their surrounding in various ways, as shown in Fig. 6.1. During their interaction, NMs can undergo various transformations which depend on their properties and environmental conditions [12]. These transformations can be classified into three groups as follows [22, 52]:

- physical transformations (aggregation and/or agglomeration, adsorption, deposition),
- chemical transformations (redox reactions, dissolution, and interaction with macromolecules), and

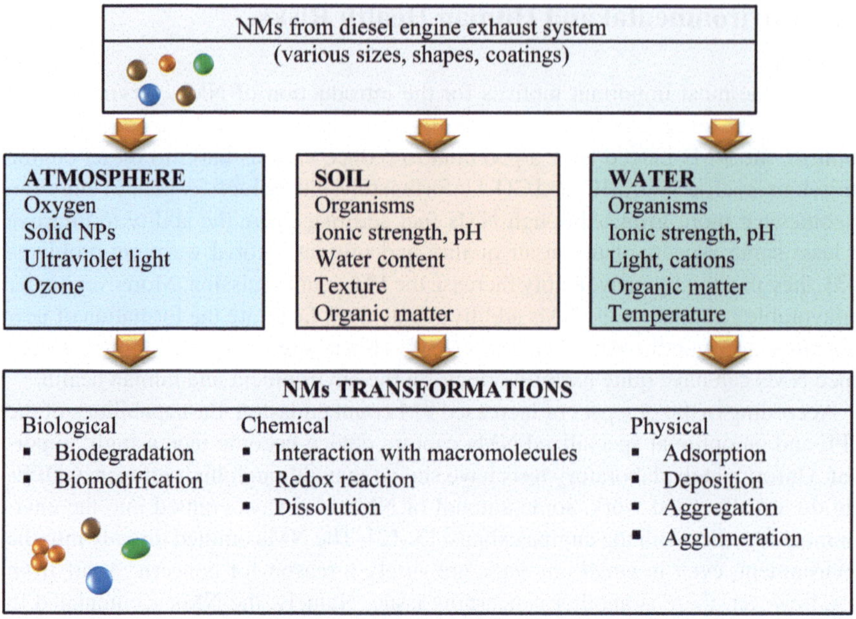

Fig. 6.1 NMs transformations and their influencing factors

- biological transformations (biodegradation and biomodifications).

The type and extent of these transformations depend mainly on the surface chemical character, surface coating, shape, and size of NMs particulates. Further influencing factors are the environmental conditions [18, 23, 35, 45, 52] of the:

- atmosphere (oxygen, solid NPs, ultraviolet light, ozone),
- soil (pH-value, water content, texture, organisms, ionic strength, organic matter), and
- water (pH-value, light, natural organic matter, organisms, ionic strength, temperature, cations).

One can say that current knowledge and understanding of the mechanisms involved in NMs interactions and transformations are far from being even close to satisfactory. From what we know, however, it seems that NMs released into the natural environment by nanofuel usage can potentially have serious negative long-term effects on atmospheric, terrestrial, and aqueous organisms.

Atmospheric Environment

Once the NMs from the engine exhaust are released into the atmosphere, they begin interacting with their surrounding, which may be accompanied by their transformations. Depending on their properties, they may tend to aggregate, agglomerate, or undergo redox reactions [52]. Either in the original form or aggregated/agglomerated into larger particles, they act as cloud condensation nuclei and promote the formation

Fig. 6.2 Environmental impacts of NMs in atmosphere

of dust clouds [18]. This may degrade the quality of air and influence light absorption and scattering, which affects the global warming processes [16]. Furthermore, organisms in the air can also interact with NMs and cause their biological transformation. For example, Zhang et al. [52] reported that iron-oxidizing bacteria can accelerate zero-valent iron corrosion by converting Fe^{2+} to Fe^{3+}.

Airborne NMs can have adverse health effects on various organisms, as shown in Fig. 6.2. The most problematic are consequences of exposures to atmospheric NMs via inhalation, skin, plant leaves, and cell membranes. For example, laboratory tests indicate that leaves of various plants can uptake and capture airborne NMs. This can cause plant damages, which may negatively affect human and animal food safety [24].

Terrestrial Environment

NMs can enter terrestrial environment from the atmosphere, e.g., by rainfall or sedimentation, and from other sources, like, fallen and contaminated plant leaves or improperly disposed contaminated DPFs. In soil, NMs may undergo various processes like sorption, aggregation, agglomeration, and dissolution, which affect their stability, mobility, and toxicity [36, 52]. For example, the increase of the dissolution rate in acidic soils can attribute to ZnO NPs becoming more hazardous. These processes are influenced by various soil-related parameters such as pH-value, water content, organisms, organic matter, ionic strength, and so on. For example, Zhang et al. [52] reported that clay fraction facilitates the retention of CeO_2 NPs in soils by forming aggregates. Meanwhile, organic matter promotes the availability of CeO_2 NPs by coating the particles and decreasing their aggregation; furthermore, it may also interact with ZnO NPs resulting in reduced particles aggregation.

NMs present in the soil can have adverse health effects on terrestrial organisms [25, 36]; some of the impacts are shown in Fig. 6.3. For example, Courtois et al. [4] reported that exposure to Ag NPs may change microbial biomass and diversity, decrease plant growth, and inhibit soil invertebrate reproduction; Bundschuh et al. [2] have shown that plants exposed to CeO_2 NPs in their parental generation show response in the filial generation with implications in the grains' nutrient composition, growth, and physiology. In general, NMs in soil can change the gene expression, biochemistry, physiology, and hence morphology of a plant. For example, Tripathi et al. [44] reported that ZnO NPs and C NTs have the ability to penetrate tissues of

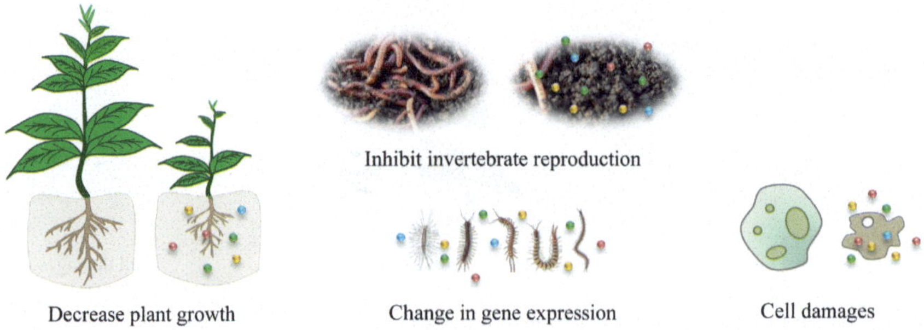

Fig. 6.3 Environmental impacts of NMs in soil

roots and seeds of *Solanum lycopersicum* by disrupting the surface of the plant; Fe_3O_4 NPs can disturb the respiration by blocking aquaporins; Ag NPs can induce oxidative stress and modify the nutritional value; CeO_2 NPs can decrease light absorption; CuO NPs can cause toxicity by entering a cell membrane and releasing Cu ions inside the cell. Furthermore, TiO_2, ZnO, NiO, and CuO NPs can produce free OH radicals, which can damage virtually all types of macromolecules; TiO_2 NPs can damage DNA and reduce transpiration rate while mwC NTs can condense chromatin inside the cytoplasm and lead to cell death. Forstner et al. [9] reported that carbon-based NMs can significantly influence the community composition of soil bacteria.

Aquatic Environment

NMs can enter aquatic environment from the atmosphere, soil, and from other sources like plant leaves. In water, NMs may undergo various processes like aggregation, agglomeration, sulfidation, and dissolution [1, 45]. These processes are influenced by various water-related parameters such as pH-value, light, ionic strength, and so on. The NMs alterations or transformations, induced by these processes, may affect MPs properties such as their toxicity. For example, Zhang et al. [52] reported that the oxidation of Ag NPs in aquatic systems notably enhances antimicrobial effects which results in higher bacterial death than observed with the original engineered Ag NPs. Furthermore, they shown that the highest toxicity of ZnO NPs in microalgae occurs in the early aging stage due to zinc ion release, while the overall toxicity decreases with time due to the formation of lesser toxic hydrozincite and zinc hydroxide. They also reported that MnO_2 NPs can undergo photo-catalyzed redox reactions to induce dissolution or generate reactive oxygen species.

NMs present in water can have adverse health effects on aquatic organisms: microbes, algae, plants, invertebrates, and vertebrates [1, 10, 45]; some of the impacts are illustrated in Fig. 6.4. Zhang et al. [52] reported that NMs can impose toxicity to microbial and algal communities mainly through changing the community compositions, diversities or activities, and decreasing biomass; especially the metallic NPs commonly affect the viability, growth, morphology, and reactive oxygen species production in bacteria and algae. Tripathi et al. [44] reported that the most important NMs impacts on microbes are: generation of reactive oxygen species, oxidative

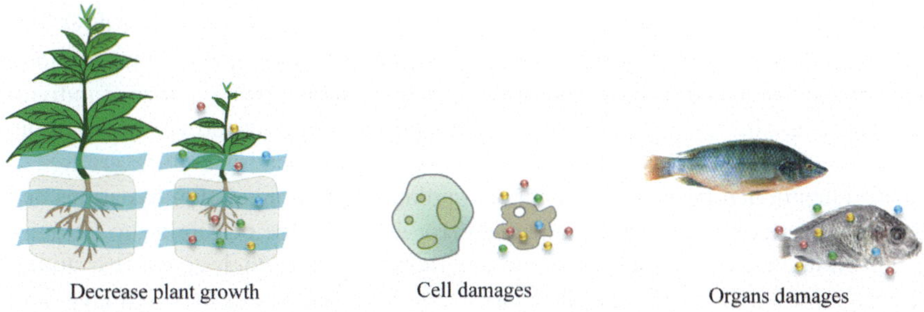

Fig. 6.4 Environmental impacts of NMs in water

damage of DNA, lesser resistance to porin synthesis, disturbances in cellular respiration, oxidative stress, and changes in the gene expression. According to Turan et al. [45], the NMs can induce a concentration-dependent increase in the reactive oxygen level species, reduction in the aquatic plant growth parameters, and the photosynthetic pigment, accompanied by an increased activity of some oxidative stress enzyme. They also reported that aquatic invertebrates exposed to NMs may suffer behavioral and physiological changes in heart rate, swimming velocity, vertical migration, and algal feeding rate. A long-term exposure of aquatic vertebrates to NMs may result in lipid peroxidation in the brain of fish, cellular damage and necrosis, and gills and liver damage [11, 45].

6.2.2 Human Health

Diesel engine emissions are one of the most important sources of pollution in urban areas and particulates are the pollutant of major concern for their effects on human health [37]. In general, most particulates from diesel engine exhaust are nanosized. These particulates may cause acute and chronic respiratory problems and are believed to be carcinogenic. Furthermore, nanosized diesel engine particulates can easily penetrate into the human lungs; they can cross cellular membranes and extend their toxicological impact on other organs. Moreover, organic particulates from a diesel engine are extremely toxic in pulmonary cell lines [13]. By using NMs-enriched fuels, the share of these problematic nanosized particulates in the diesel engine exhaust becomes even larger. This necessitates thorough studies on their impact on human health.

With regard to the biological test system, three types of studies are possible: epidemiological, in-vivo, and in-vitro [41]. Epidemiological studies, where the biological system is a specific cohort population, can only give retrospective estimates of the effects related to specific technologies that are already in widespread use; in our case, this is not feasible because NMs additives have not yet been used over longer time periods. In-vivo studies, where the biological system is a short life-time animal (mice or rat) or a human, suffer from ethical concerns. Furthermore, they are expensive and usually cannot be performed with a high number of individuals that would be necessary to overcome the problem of inter-individual variation. In-vitro studies, where the biological system is a cell or bacterial culture, are free of the aforementioned burdens, but cannot simulate the real inhalation dynamics and systemic reactions of the whole organism (for example, the possibility to detoxify certain compounds via the bloodstream and the liver). Nevertheless, in-vitro studies are a valuable and indispensable alternative to in-vivo studies due to their high potential for standardization, high efficiency, low cost, and absence of ethical concerns. Therefore, the influence of NMs fuel additives is most frequently investigated by in-vitro studies [41].

Unfortunately, recent research has raised rather serious concerns about the safety of nanofuels usage, related mainly to long-term adverse effects on human health [27]. Namely, due to their high chemical reactivity and biological activity, the NMs particles found in engine exhaust are able to cross biological membranes and access cells, tissues, and organs that larger-sized particles normally cannot. Therefore, they can become quite harmful once they enter the bloodstream via penetration through skin, inhalation, or ingestion of water and food, which leads to both, genotoxicity and biochemical toxicity [33, 44, 45]. Clearly, the toxicity of a particular NMs depends on its type. Besides this, however, the harmful effects on human health depend also on the dosage, exposure time, aggregation and concentration effects, size, shape, surface area, and crystal structure of NMs [17]. For example, the CeO_2 NPs, frequently tested as fuel additive, were investigated with respect to their effect on organ damage and diseases by Cassee et al. [3]. They reported that the size and shape of CeO_2 NPs determine the solubility, and thereby the possibility of the NMs uptake in the cells of the respiratory tract.

Under the bottom line, it should be obvious that human exposure to nanofuel-borne NMs can potentially lead to a number of harmful effects. According to current knowledge, these effects include organ damages and diseases, cellular and genetic effects, and biochemical changes [27, 33], Fig. 6.5.

Organ Damage and Diseases

As recent studies clearly indicate, NMs can accumulate in various organs, especially in brain, lungs, and liver, where they may cause various health issues. Hochella et al. [16] investigated metallic NMs and found out that these can enter the brain directly through the olfactory bulb. For example, metallic Fe_3O_4 NPs were found in human brain tissue; these NPs are believed to have potentially negative effects on brain health. Guo et al. [15] investigated the pulmonary toxicity of CeO_2 NPs of the size characteristic for exhaust gasses from a diesel engine. From the study, conducted on test animals, it follows that the markers of inflammation may change significantly in a dosage- and time-dependent manner. In all investigated cases, the inflammatory response was greater than that observed in studies by using micro-sized ceria aerosols. Jeevanandam et al. [17] investigated the impacts of various metallic NMs; they reported that these NMs have the capability to damage lung tissues by

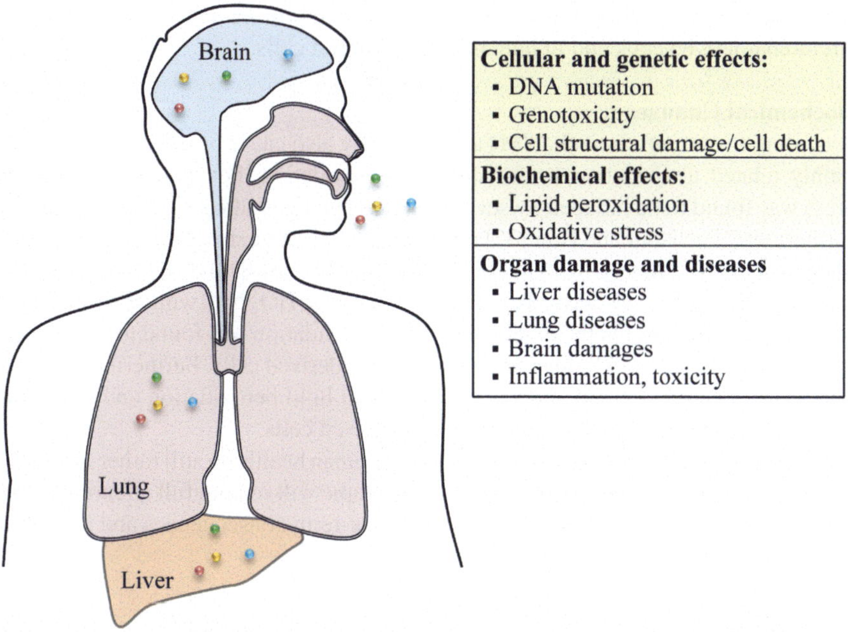

Fig. 6.5 Human health impacts of NMs

producing reactive oxygen species. Other studies have found that addition of iron NPs to mineral diesel may increase inflammatory effects [42] while the addition of Ag, Fe^0 and SiO_2 NPs and swCN NTs, can induce inflammation and pulmonary toxicity [32, 34]. Nalabotu et al. [29] reported that intratracheal instillation of CeO_2 NPs induced hepatic toxicity in test animals, due to traveling of CeO_2 NPs from the lungs to the liver, causing increased liver ceria levels, reduction in liver weight, and evidence of liver damage.

Cellular and Genetic Effects

Unfortunately, current research indicates that NMs additives may have various genetic effects. For example, Steiner et al. [42] reported that the addition of iron NPs to low-sulfur mineral diesel fuel seems to decrease mutagenic potentials. Furthermore, Zhang and Balasubramanian [51] have found that exposure of human epithelial cells to diesel particulates can cause alterations in the global gene expression with a broad range of biochemical pathways. This can be attributed to the changes in the physicochemical characteristics of diesel particulates, caused by the NMs additives. Pietroiusti et al. [34] reported that several metallic and metallic oxide NMs seem to have the ability to induce genotoxicity. Besides genetic effects, nanofuels may exhibit a potential to increase cytotoxicity of emitted particulates. This was observed by Pachapur and Brar [32] when using Ag NPs in tests with human macrophage and lung fibroblasts; similar observations have been made when using SiO_2 NPs with human bronchoallveolar-derived cells. Furthermore, Kabir et al. [18] reported that ZnO NPs may reduce cell viability and increase DNA damage in human bronchoallveolar carcinoma-derived cells and in human lung fibroblast cells.

Biochemical Changes

According to current research, NMs additives may also cause biochemical changes, mainly related to oxidative stress and lipid peroxidation. For example, oxidative stress was found to be induced by several metallic and metallic oxide NMs [34, 42]. Furthermore, Pachapur and Brar [32] reported that oxidative stress was also observed when using C NTs in tests with human cell lines, when using SiO_2 NPs in tests with human bronchoallveolar-derived cells, and when using TiO_2 NPs with bone-marrow-derived macrophages. On the other hand, lipid peroxidation was found to be induced by SiO_2 NPs, tested with human bronchoallveolar-derived cells. Furthermore, Kabir et al. [18] reported that ZnO NPs may cause both lipid peroxidation and oxidative stress in human bronchoallveolar carcinoma-derived cells.

Unfortunately, the impacts of NMs usage on human health are still rather modestly investigated. It looks obvious therefore that this topic will require full attention in the future in order to prevent possibly damaging long-term consequences and to enable safe engagement of valuable NMs fuel additives.

References

1. Amde M, Liu JF, Tan ZQ, Bekana D (2017) Transformation and biovailability of metal oxide nanoparticles in aquatic and terrestrial environments. A review. Env Pollut 230:250–267
2. Bundschuh M, Filser J, Lüderwald S, McKee MS, Metreveli G, Schaumann GE, Schulz R, Wagner S (2019) Nanoparticles in the environment: where do we come from, where do we go to? Environ Sci Eur 30:6
3. Cassee FR, vanBalen EC, Singh C, Green D, Muijser H, Weinstein J, Dreher K (2011) Exposure, health and ecological effects review of engineered nanoscale cerium and cerium oxide associated with its use as a fuel additive. Crit Rev Toxicol 41:213–229
4. Courtois P, Rorat A, Lemiere S, Guyoneaud R, Attard E, Levard C, Vandenbulcke F (2019) Ecotoxicology of silver nanoparticles and their derivatives introduced in soil with or without sewage sludge: a review of effects on microorganisms, plants and animals. Environ Pollut 253:578–598
5. Dale JG, Cox SS, Vance ME, Marr LC, Hochella MF (2017) Transformation of Cerium oxide nanoparticles from a diesel fuel additive during combustion in a diesel engine. Environ Sci Technol 51:1973–1980
6. Damanik N, Ong HC, Tong CW, Mahlia TMI, Silitonga AS (2018) A review on the engine performance and exhaust emission characteristics of diesel engines fueled with biodiesel blends. Environ Sci Pollut Res 25(16):15307–15325
7. El-Seesy AI, Attia AMA, El-Battsh M (2018) The effect of Aluminum oxide nanoparticles addition with Jojoba methyl ester-diesel fuel blend on a diesel engine performance, combustion and emission characteristics. Fuel 224:147–166
8. Fang J, Meng Z, Li J, Pu Y, Du Y, Li J, Jin Z, Chen C, Chase GG (2017) The influence of ash on soot deposition and regeneration processes in diesel particulate filter. Appl Therm Eng 124:633–640
9. Forstner C, Orton TG, Wang P, Kopittke PM, Dennis PG (2019) Effects of carbon nanotubes and derivatives of graphene oxide on soil bacterial diversity. Sci Total Environ 682:356–464

10. Freixa A, Acuña V, Sanchís J, Farré M, Barceló D, Sabater S (2018) Ecotoxicological effects of carbon based nanomaterials in aquatic organisms. Sci Total Environ 619–620:328–337
11. Ghais SA, Bhardwaj V, Kumbhar P, Shehhi OA (2019) Effect of copper nanoparticles and organometallic compounds (dibutyltin) on tilapia fish. J Basic Appl Zool 80:32
12. Giese B, Klaessig F, Park B, Kaegi R, Steinfeldt M, Wigger H, Gleich A, Gottschalk F (2018) Risks, release and concentrations of engineered nanomaterial in the environment. Sci Rep 8:1565
13. Gualtieri M, Capasso L, D'Anna A, Camtini M (2014) Organic nanoparticles from different fuel blends: in vitro and inflammatory potential. J Appl Toxicol 34:1247–1255
14. Guan B, Zhan R, Lin H, Huang Z (2015) Review of the state-of-the-art of exhaust particulate filter technology in internal combustion engines. J Environ Manage 154:225–258
15. Guo C, Robertson S, Weber RJM, Buckley A, Warren J, Hodgson A (2019) Pulmonary toxicity of inhaled nano-sized cerium oxide aerosols in Sprague-Dawley rats. Nanotoxicology 13:733–750
16. Hochella MF, Mogk DW, Ranville J, Allen IC, Luther GW, Marr LC, McGrail BP, Murayama M, Qafoku NP, Rosso KM, Sahai N, Schroeder PA, Vikesland P, Westerhoff P, Yang Y (2019) Natural, incidental, and engineered nanomaterials and their impacts on the Earth system. Science 363:1414
17. Jeevanandam J, Barhoum A, Chan YS, Dufresne A, Danquah MK (2018) Review on nanoparticles and nanostructured materials: history, sources, toxicity and regulations. Beilstein J Nanotech 9:1050–1074
18. Kabir E, Kumar V, Kim KH, Yip ACK, Sohn JR (2018) Environmental impacts of nanomaterials. J Environ Manage 225:261–271
19. Khond VW, Kriplani VM (2016) Effect of nanofluid additives on performances and emissions of emulsified diesel and biodiesel fueled stationary CI engine: a comprehensive review. Renew Sustain Energy Rev 59:1338–1348
20. Kumar MV, Babu AV, Kumar PR (2019) Influence of metal-based cerium oxide nanoparticle additive on performance, combustion, and emissions with biodiesel in diesel engine. Environ Sci Pollut Res 26:7651–7664
21. Kumar S, Dinesha P, Bran I (2017) Influence of nanoparticles on the performance and emission characteristics of a biodiesel fuelled engine: an experimental analysis. Energy 140:98–105
22. Lei C, Sun Y, Tsang DCW, Lin D (2018) Environmental transformations and ecological effects of iron-based nanoparticles. Environ Pollut 232:10–30
23. Lowry GV, Avellan A, Gilbertson LM (2019) Opportunities and challenges for nanotechnology in the agri-tech revolution. Nat Nanotechnol 14:517–522
24. Lv J, Christie P, Zhang S (2019) Uptake, translocation, and transformation of metal-based nanoparticles in plants: recent advances and methodological challenges. Environ Sci Nano 6:41–59
25. McKee MS, Filser J (2016) Impacts of metal-based engineered nanomaterials on soil communities. Environ Sci Nano 3:506–533
26. Meesters JAJ, Quik JTK, Koelmans AA, Hendriks AJ, van de Meent D (2016) Multimedia environmental fate and speciation of engineered nanoparticles: a probabilistic modeling approach. Environ Sci Nano 3:715–727
27. Missaou WN, Arnold RD, Cummnings BS (2018) Toxicological status of nanoparticles: what we know and what we don't know. Chem Biol Interact 295:1–12
28. Mohankumar S, Senthilkumar P (2017) Particulate matter formation and its control methodologies for diesel engine: a comprehensive review. Renew Sustain Energy Rev 80:1227–1238
29. Nalabotu SK, Kolli MB, Triest WE, Ma JY, Manne NDPK, Katta A, Addagaria HS, Rice KM, Blough ER (2011) Intratracheal instillation of cerium oxide nanoparticles induces hepatic toxicity in male Sprague-dawles rats. Int J Nanomed 6:2327–2335
30. Nour M, El-Seesy AI, Abdel-Rahman AK, Bady M (2018) Influence of adding aluminum oxide nanoparticles to diesterol blends on the combustion and exhaust emission characteristics of a diesel engine. Exp Thermal Fluid Sci 98:634–644

31. Orihuela MP, Gómez-Martín A, Miceli P, Becerra JA, Chacartegui R, Fino D (2018) Experimental measurement of the filtration efficiency and pressure drop of wall-flow diesel particulate filters (DPF) made of biomorphic silicon carbide using laboratory generated particles. Appl Therm Eng 131:41–53
32. Pachapur V, Brar SK (2016) Nano-ecotoxicology of natural and engineered nanomaterials for animals and humans. In: Zhang TC, Verma M, Surampalli RY, Tyagi RD (eds) Brar SK. Nanomater Env ASCE, Reston
33. Patil SS, Shedbalkar UU, Truskewycz A, Chopade BA, Ball AS (2016) Nanoparticles for environmental clean-up: a review of potential risks and emerging solutions. Environ Technol Innov 5:10–21
34. Pietroiusti A, Stocmann-Juvala H, Lucaroni F, Savolainen K (2018) Nanomaterial exposure, toxicity, and impact on human health. Wires Nanomed Nanobiotechnol 10:1513
35. Prajitha N, Athira SS, Mohanan PV (2019) Bio-interactions and risks of engineered nanoparticles. Environ Res 172:98–108
36. Rajput V, Minkina T, Sushkova S, Behal A, Maksimov A, Blicharska E, Ghazaryan K, Movsesyan H, Barsova N (2019) ZnO and CuO nanoparticles: a treat to soil organisms, plants, and human health. Environ Geochem Health. https://doi.org/10.1007/s10653-019-00317-3
37. Rehna VJ, Siddique A (2018) Risk evaluation and exposure hazards of engineered nanomaterials: a survey. Am J Eng Res 7(1):235–245
38. Saxena V, Kumar N, Saxena VK (2017) A comprehensive review on combustion and stability aspects of metal nanoparticles and its additive effect on diesel and biodiesel fueled CI engine. Renew Sustain Energy Rev 70:563–588
39. Shaafi T, Sairam K, Gopinath A, Kumaresan G, Velrajm R (2015) Effect of dispersion of various nanoadditives on the performance and emission characteristics of a CI engine fueled with diesel, biodiesel and blends—a review. Renew Sustain Energy Rev 49:563–573
40. Soudagar MEM, Nik-Ghazali NN, Kalam MA, Badruddin IA, Banapurmath NR, Akram N (2018) The effect of nano-additives in diesel-biodiesel fuel blends: a comprehensive review on stability, engine performance and emission characteristics. Energy Convers Manag 178:146–177
41. Steiner S, Bisig C, Petri-Fink A, Rothen-Rutishauser B (2016) Diesel exhaust: current knowledge of adverse effects and underlying cellular mechanisms. Arch Toxicol 90:1541–1553
42. Steiner S, Czerwinski J, Comte P, Heeb NV, Mayer A, Petri-Fink A, Rothen-Rutishauser B (2015) Effects of an iron-based fuel-borne catalyst and a diesel particle filter on exhaust toxicity in lung cells in vitro. Anal Bioanal Chem 407:5977–5986
43. Tollefson J (2013) Soot a major contributor to climate change. Nature. https://doi.org/10.1038/nature.2013.12225
44. Tripathi DK, Ahmad P, Sharma S, Chauchan DK, Dubey NK (2017) Nanomaterilas in plants, algae, and microorganisms. Elsevier, London
45. Turan NB, Erkan HS, Engin GO, Bilgili MS (2019) Nanoparticles in the aquatic environment: usage, properties, transformation and toxicity—a review. Process Saf Environ Prot 130:238–249
46. Vellaiyan S, Subbiah A, Chockalingam P (2019) Multi-response optimization to improve the performance and emissions level of a diesel engine fueled with ZnO incorporated water emulsified soybean biodiesel/diesel fuel blends. Fuel 237:1013–1020
47. Venu H, Raju VD, Subramani L (2019) Combined effect of influence of nano additives, combustion chamber geometry and injection timing in a DI diesel engine fuelled with ternary (diesel-biodiesel-ethanol) blends. Energy 174:386–406
48. Wu Q, Xie X, Wang Y, Roskilly T (2018) Effect of carbon coated aluminum nanoparticles as additive to biodiesel-diesel blends on performance and emission characteristics of diesel engine. Appl Energy 221:597–604
49. Yaşar A, Keskin A, Yıldızhan Ş, Uludamar E (2019) Emission and vibration analysis of diesel engine fueled diesel fuel containing metallic based nanoparticles. Fuel 239:1224–1230

50. Zhang ZH, Balasubramanian R (2015) Influence of an iron-based fuel-borne catalyst on physicochemical and toxicological characteristics of particulate emissions from a diesel engine. Appl Energy 146:270–278
51. Zhang ZH, Balasubramanian R (2017) Effects of cerium oxide and ferrocene nanoparticles addition as fuel-borne catalysts on diesel engine particulate emissions: environmental and health implications. Environ Sci Technol 51:4248–4258
52. Zhang J, Guo W, Li Q, Wang Z, Liu S (2018) The effects and the potential mechanism of environmental transformation of metal nanoparticles on their toxicity in organisms. Environ Sci Nano 5:2482–2499

Chapter 7
Conclusions and Future Perspectives

According to recent research, NMs fuel additives exhibit a promising potential to improve various aspects of diesel engines. To provide more insight into this exciting and challenging development, this book offers a comprehensive review on recent investigations related to NMs-enriched fuels usage in diesel engines. Besides the carefully elaborated short-term results, mostly related to diesel engine, the potential long-term consequences, which include environmental and human health aspects, are also addressed briefly. The topic considered is interdisciplinary and extremely complex, making it very difficult to extract and formulate concise and undeniable conclusions. Nevertheless, an attempt is made in the following to provide a summary of valuable concluding remarks and an outline of perhaps the most needed activities in the near future.

Base fuels. The most frequently tested base fuels (to be enriched with NMs) are neat mineral diesel, neat biodiesels, and blends of mineral diesel and up to 25% biodiesel. Other fuels, tested to a lesser extent, are: neat gas-to-liquid fuel, blends of mineral diesel and various alcohols, blends of mineral diesel, various biodiesels and various alcohols, blends of biodiesel and oil, and water-in-various fuels emulsions.

Nanomaterials. The most frequently tested NMs engaged as fuel additives are Al_2O_3 NPs, CeO_2 NPs, CuO NPs, TiO_2 NPs, ZnO NPs, and carbon NMs. Mostly, these NMs are synthesized by the sol-gel, sol-gel combustion, hydrothermal, co-precipitation, electric arc discharge, and mechanical ball milling methods, due to their either simplicity, or efficiency, or environmental friendliness. Typical size of the synthesized NMs is roughly about 50 nm or less.

Nanofuels preparation. Most of the tested nanofuels were prepared by two-step methods, where either NMs powder or NMs nanofluid is dispersed (typically by the ultrasonication method) into the base fuel during the post-production process.

Nanofuels stability. The prepared nanofuels are in general thermodynamically unstable systems. The stability can be improved by adapting the preparation methods and by adding various surfactants, such as 2-ethylhexyl nitrate, Darvan-C, CTAB,

T. Kegl et al., *Nanomaterials for Environmental Application*, Green Energy and Technology, https://doi.org/10.1007/978-3-030-54708-0_7

sorbitan oleate, Span 80™, Tween 80™, oleic acid, citric acid, EtOH, isopropanol, TMAOH, and SDS.

Nanofuels properties. Although the nanofuel properties depend to a great extent on NMs type, size, and dosage, the addition of NMs to some base fuel typically increases the fuel density, cetane number, and to some extent also the calorific value. A variation in fuel density influences heavily the injection process characteristics in a MC-FIS and consequently indirectly also the combustion process, emissions, and engine performance. Meanwhile, a variation of the cetane number and calorific value affects directly the combustion process, emissions, and engine performance.

NMs activities in a diesel engine. The most important NMs activities can be classified into four groups: (i) secondary atomization due to disruption behavior and micro-explosions, especially emphasized in fuels consisting of multiple components with different boiling points, (ii) catalytic engagement in various reactions during the combustion process or after-treatment of combustion products, (iii) reactivity engagement either directly or in oxidation or reduction with either some fuel component or combustion products, and (iv) thermal engagement, typically by acting as a heat sink, capable of absorbing heat and reducing the in-cylinder temperature.

NMs transformations in a diesel engine. NMs properties may change significantly during their lifetime in a diesel engine due to already known as well as yet unknown transformations. Most notably, some NMs may dissociate at high temperature during the combustion process, while some NMs may change their morphology and size or other properties.

General nanofuel effects on diesel engine characteristics. The most important engine characteristics depend on the engaged NMs type and their properties, NMs size and dosage, base fuel, nanofuel preparation technique, injection system type, engine type, and instantaneous engine operating regime.

- **NMs effects on injection characteristics**: The presence of NMs can influence significantly the injection characteristics (e.g., injection timing, injection pressure) in a MC-FIS due to induced fuel density variation. Meanwhile, a CR-FIS is not affected, because it is controlled electronically and can be adapted to fuel density variation.
- **NMs effects on fuel spray characteristics**: NMs addition can influence the macroscopic and microscopic characteristics of the fuel spray. With respect to the base fuel, a nanofuel spray usually exhibits a wider spray angle and shorter liquid spray breakup length due to reduced surface tension and increased fuel viscosity as well as smaller mean droplet diameters due to secondary atomization.
- **NMs effects on combustion characteristics**: NMs addition to the base fuel typically reduces the ignition delay due to the higher cetane number and thermal conductivity of the nanofuel, and due to the NMs catalytic engagement. The in-cylinder pressure can increase due to the higher cetane number and lower viscosity of the nanofuel, and NMs catalytic and reactive engagements. Furthermore, the in-cylinder pressure can also decrease due to higher evaporation rate of

the nanofuel. The heat release rate can increase due to higher calorific value and thermal conductivity of the nanofuel and due to catalytic activity of NMs. On the other hand, the heat release rate can decrease due to the thermal activity of NMs, acting as a heat sink. Both, in-cylinder temperature and exhaust gas temperature, can either increase due to the higher calorific value of the nanofuel or decrease due to the thermal activity of NMs acting as a heat sink.

- **NMs effects on engine emissions**: NO_x, HC, CO, and PM emissions depend mainly on the base fuel type, NMs activities (secondary atomization, catalytic activity …), and nanofuel properties (cetane number, oxygen content). Most of the experimental studies show that NMs addition to the base fuel practically unavoidably increases the PM count emissions. NO_x emission is increased in about one-third of reported tests, while the rest of the tests indicate a slight or sometimes even a significant NO_x emission decrease. Meanwhile, CO, unburned HC, smoke and PM mass emissions typically decrease by NMs additives engagement. Under certain special conditions, however, just the opposite effects are also possible.
- **NMs effects on engine performance**: NMs addition to base fuels usually results in reduced brake specific fuel consumption and higher brake thermal efficiency. The most exposed reasons are NMs catalytic and reactive activity, promoted secondary atomization, as well as higher cetane number and thermal conductivity of the nanofuel, compared to the base fuel.

NMs effects on diesel engine parts and systems. NMs can potentially have an important role in preventing carbon, metals, and other materials to deposit on cylinder walls, which could significantly decrease friction. The altered physical and chemical properties of NMs-enriched fuels require a modified nozzle geometry, a modified piston geometry in order to ensure an adequate swirl in the combustion chamber, and modified programming/maps of the ECU.

NMs risks. Unfortunately, NMs might have potential toxic effects on living organisms. The underlying investigations, being still rather scarce at the time, indicate that the engagement of NMs fuel additives in diesel engines should raise concerns about their potential environment and human health risks. For example, ZnO NPs exhibit high cytotoxicity in terms of cell proliferation, cell viability, and membrane integrity. Furthermore, TiO_2 NPs can damage DNA and reduce transpiration rate, while mwC NTs can condense chromatin inside the cytoplasm and cause cell death. Some studies have found that addition of iron NPs to mineral diesel fuel can increase inflammatory effects, while CeO_2 NPs can cause membrane damage.

The introduction of NMs-enriched fuels into diesel engines is a complex and sophisticated undertaken, plagued by conflicting situations, deficient knowledge, and difficult challenges. Finding the right solutions and proper balance between pros and contras will determine the viability of NMs fuel additives and the ultimate fate of nanofuels usage. In order to make this a success story, the nanofuels have to be made both attractive and acceptable in the short- and long-term usage from the technical, environmental, and human health aspects.

From the technical aspect, the following topics should probably get the most attention:

- **Determination of optimal NMs size and dosage**. This will require extensive research work due to vast abundance of possible base fuels and NMs types. Once successful, however, this would provide attractive engine performance and emissions, which would provide a solid foundation for further work on other technical and environmental issues.
- **Development of suitable NMs synthesis methods**. Once an attractive NMs type and its optimal size are determined, a suitable synthesis method will be needed which has to deliver the particles of required size in a technologically and economically viable way.
- **Development of nanofuel preparation methods and systems**. In case of stored nanofuels, suitable preparation methods and optional surfactants have to be found to assure long-term stable, homogeneous, and aggregation/agglomeration-free suspensions. If the nanofuel has to be prepared just before the delivery into the cylinder, a suitable device has to be developed and interfaced to the ECU to enable adjustable NMs dosage in dependence on engine operating regime.
- **Development of advanced DPFs**. As it looks, it will be absolutely necessary to minimize the release of NMs into the environment to negligible levels. This will require much more sophisticated DPFs, preferably with a possibility to extract and possibly recycle NMs particulates from the exhaust system.

From the environmental and human health aspects, the following topics are probably the most important:

- **Study of transformations of NMs-related particulates**. After the engineered NMs enter the engine cylinder, they may transform in various ways: firstly, during the short but violent lifetime in the diesel engine and exhaust system, and secondly, during their lifetime in the atmosphere, soil, and aqueous systems. This knowledge is absolutely necessary since it will provide the foundation to study the actual environmental impact of the released particulates.
- **Study of short- and long-term NMs effects on atmospheric, terrestrial, and aquatic environments and organisms**. This topic, especially the NMs impacts on human health, is of prominent importance because one cannot realistically expect that DPFs or other similar devices, even if being very advanced, will ever capture all of the NMs particulates entering the engine exhaust system.

Already a brief glance at these topics reveals the amount and complexity of work needed to make diesel engines running on nanofuels an attractive and viable option. It seems, however, that NMs fuel additives are worth the effort since the potential reward might be a much cleaner diesel engine. This is especially important because, when looking closely, diesel engine alternatives are also plagued by various issues and problematic environmental impacts when taking into account the whole lifecycle. This corroborates the impression that diesel engines will remain important and much needed machines for quite a long time.

The manufacturer's authorised representative in the EU is Springer Nature Customer Service Centre GmbH, Europaplatz 3, 69115 Heidelberg, Germany. If you have any concerns regarding our products, please contact ProductSafety@springernature.com

Printed and bound by CPI Group (UK) Ltd, Croydon, CR0 4YY
30/09/2026
02225504-0001